Soil Behavior and Geomechanics

Soil Behavior and Geomechanics

Contributors

Wisley Moreira Farias, Geraldo Resende Boaventura et al.

AURIS
Reference

www.aurisreference.com

Soil Behavior and Geomechanics

Contributors: Wisley Moreira Farias, Geraldo Resende Boaventura et al.

Published by Auris Reference Limited

www.aurisreference.com

United Kingdom

Soil Behavior and Geomechanics

ISBN: 978-1-78154-837-0

British Library Cataloguing in Publication Data
A CIP record for this book is available from the British Library

Printed in the United Kingdom

Exclusively distributed by CBS Publishers & Distributors Pvt. Ltd.

Sales & Distribution Rights only for India, Pakistan, Bangladesh, Sri Lanka, Nepal and Bhutan.This book is not to be sold outside these territories.

Contents

List of Abbreviations

AES	Atomic Emission Spectrometry
CBR	Californian Bearing Ration
CEC	Cation Exchange Capacity
CRM	Certified Reference Materials
DCB	Dithionite-Citrate Bicarbonate
DSC	Differential Scanning Calorimetric
DT	Differential Thermal
EC	European Community
ECBR	European Community Bureau of Reference
EDS	Energy Dispersive Spectra
FEM	Finite Elements Model
FTIR	Fourier Transform Infra-Red
GPS	Global Positioning System
HCA	Hierarchical Cluster Analysis
LEFM	Linear Elastic Fracture Mechanics
LNALP	Light Non-Aqueous Liquid Phase
MB	Methylene Blue
MPAC	Maximum Phosphate Adsorption Capacity
NPP	Net Primary Productivity
OM	Organic Matter
OPM	Optimum Modified Proctor
PC	Principal Component
PCA	Principal Component Analysis
POP	Persistent Organic Pollutant
PSD	Particle Size Distribution
RAC	Risk Assessment Code
RCC	Rotating Coiled Columns
SEM	Scanning Electron Microscope
SEP	Sequential Extraction Procedure
SOC	Soil Organic Carbon
SQG	Sediment Quality Guideline
TDR	Time Domain Reflectometry
TG	Thermo-Gravimetric
TGA	Thermo-Gravimetric Analysis
TL	Thin Lamina
UCT	Unconfined Compression Test
USDA	United States Department of Agriculture
WAS	West African Standards
XRD	X-Ray Diffraction

List of Contributors

Moreira Farias
Universidade de Brasília, Brasília-DF, Brazil

Geraldo Resende Boaventura
Universidade de Brasília, Brasília-DF, Brazil

Éder de Souza Martins
Embrapa/ Cerrados, Planaltina –DF, Brazil

Fabrício Bueno da Fonseca Cardoso
ANA-Agência Nacional de Águas –Brasília-DF, Brazil

José Camapum de Carvalho
Universidade de Brasília, Brasília-DF, Brazil

Edi Mendes Guimarães
Universidade de Brasília, Brasília-DF, Brazil

Rainer Schuhmann
Karlsruhe Institute of Technology (KIT) Competence Centre for Material Moisture (CMM) Karlsruhe, Germany

Franz Königer
Karlsruhe Institute of Technology (KIT) Competence Centre for Material Moisture (CMM) Karlsruhe, Germany

Katja Emmerich
Karlsruhe Institute of Technology (KIT) Competence Centre for Material Moisture (CMM) Karlsruhe, Germany

Eduard Stefanescu
Karlsruhe Institute of Technology (KIT) Competence Centre for Material Moisture (CMM) Karlsruhe, Germany

Markus Stacheder
Karlsruhe Institute of Technology (KIT) Competence Centre for Material Moisture (CMM) Karlsruhe, Germany

Elisabeth Olivia Logmo
Laboratory of Subsurface, Department of Earth Sciences, Faculty of Science, University of Douala, Douala, Cameroon

Gilbert François Ngon Ngon
Laboratory of Subsurface, Department of Earth Sciences, Faculty of Science, University of Douala, Douala, Cameroon

Williams Samba
Laboratory of Subsurface, Department of Earth Sciences, Faculty of Science, University of Douala, Douala, Cameroon

Michel Bertrand Mbog
Laboratory of Environmental Geology, Department of Earth Sciences, Faculty of Science, University of Dschang, Dschang, Cameroon

Jacques Etame
Laboratory of Subsurface, Department of Earth Sciences, Faculty of Science, University of Douala, Douala, Cameroon

Meissa Fall
Laboratoire de Mécanique et Modélisation, UFR Sciences de l'Ingénieur, University of Thies, Thies, Senegal

Déthiè Sarr
Laboratoire de Mécanique et Modélisation, UFR Sciences de l'Ingénieur, University of Thies, Thies, Senegal

Makhaly Ba
Laboratoire de Mécanique et Modélisation, UFR Sciences de l'Ingénieur, University of Thies, Thies, Senegal
Département de Génie Civil, UFR SI-Université de Thiès, Thiès, Sénégal

Etienne Berbinau
RAZEL sa, Christ de Sarclay, 3 rue René razel, Orsay cedex, Orsay, France

Jean-Louis Borel
RAZEL sa, Christ de Sarclay, 3 rue René razel, Orsay cedex, Orsay, France

Mapathé Ndiaye
Laboratoire de Mécanique et Modélisation, UFR Sciences de l'Ingénieur, University of Thies, Thies, Senegal
Département de Génie Civil, UFR SI-Université de Thiès, Thiès, Sénégal

Cheikh H. Kane
Laboratoire de Mécanique et Modélisation, UFR Sciences de l'Ingénieur, University of Thies, Thies, Senegal

Santosh Kumar Sarkar
Department of Marine Science, University of Calcutta, Calcutta, West Bengal, India

Paulo J.C. Favas
Department of Geology, School of Life Sciences and the Environment, University of Trás-os-Montes e Alto Douro, Vila Real, Portugal
IMAR-CMA Marine and Environmental Research Centre, Faculty of Sciences and Technology, University of Coimbra, Coimbra, Portugal

Dibyendu Rakshit
Department of Marine Science, University of Calcutta, Calcutta, West Bengal, India

K.K. Satpathy
Indira Gandhi Centre for Atomic Research, Environment and Safety Division, Kalpakkam, Tamil Nadu, India

Valdinar Ferreira Melo
Department of Soil and Agricultural Engineering, Federal University of Roraima, Boa Vista, Brazil

Sandra Cátia Pereira Uchôa
Department of Soil and Agricultural Engineering, Federal University of Roraima, Boa Vista, Brazil

Zachary N. Senwo
Department of Biological & Environmental Sciences, Alabama A&M University, Huntsville, USA

Ronilson José Pedroso Amorim
Agronomy, Federal University of Roraima, Boa Vista, Brazil

Musab Aied Qissab
Department of Civil Engineering, Al-Nahrain University, Baghdad, Iraq

Oustasse Abdoulaye Sall
Département de Génie Civil, UFR SI-Université de Thiès, Thiès, Sénégal

Daouda Sangare
Département de Mathématiques, UFR SAT-Université Gaston Berger, Saint Louis, Sénégal

Mathioro Fall
Département de Génie Civil, UFR SI-Université de Thiès, Thiès, Sénégal

Alassane Thiam
Département de Génie Civil, UFR SI-Université de Thiès, Thiès, Sénégal

Jian Li
Department of Geosciences and Environmental Engineering, Southwest Jiaotong University, Chengdu, Sichuan 610031, China

Xiyong Wu
Department of Geosciences and Environmental Engineering, Southwest Jiaotong University, Chengdu, Sichuan 610031, China

Long Hou
Wanzhou District Commission of Urban-Rural Development, Wanzhou, Chongqing 404000, China

Siul Ruiz
Department of Environmental Systems Science, ETHZ, Zurich, Switzerland

Dani Or
Department of Environmental Systems Science, ETHZ, Zurich, Switzerland

J. Schymanski
Department of Environmental Systems Science, ETHZ, Zurich, Switzerland

Bang-lin Luo
College of Resources and Environment/Key Laboratory of Eco-environment in Three Gorges Region (Ministry of Education), Southwest University, Chongqing, China

Xiao-yan Chen
College of Resources and Environment/Key Laboratory of Eco-environment in Three Gorges Region (Ministry of Education), Southwest University, Chongqing, China

Lin-qiao Ding
College of Resources and Environment/Key Laboratory of Eco-environment in Three Gorges Region (Ministry of Education), Southwest University, Chongqing, China

Yu-han Huang
College of Resources and Environment/Key Laboratory of Eco-environment in Three Gorges Region (Ministry of Education), Southwest University, Chongqing, China

Ji Zhou
State Key Lab of Urban and Regional Ecology, Research Centre for Eco-Environmental Sciences, Chinese Academy of Sciences, Beijing, China

Tian-tian Yang
Department of Civil and Environmental Engineering, University of California Irvine, Irvine, California, United States of America

Preface

The book Soil Behavior and Geomechanics covers recent advances in unsaturated soil mechanics, experimental study of soil behavior, and constitutive, numerical, and multi-scale modeling of soil behavior. First chapter focuses on chemical and hydraulic behavior of a tropical soil compacted submitted to the flow of gasoline hydrocarbons. In second chapter, we will focus on the important properties relevant to the solid matrix of fine grained soils which include the texture and fabric and its mineral phase content. The objective of third chapter is to associate the geotechnical characteristic to the mineralogical and chemical compositions of the clay occurrences of the Missole II deposit in order to evaluate its suitability for manufacturing of construction materials and ceramics. Fourth chapter is primarily intended to demonstrate that under unpredicted traffic and repeated loading, properties of gravel lateritic soils used as pavement layer can significantly change. Fifth chapter aims to summarize the potentials of sequential extraction technique adopting different analytical protocols for gaining information on the mobility and dynamics of operationally determined chemical forms of heavy metals in soils and sediments. Sixth chapter evaluate the maximum phosphate adsorption capacity (MPAC) of the soils developed from mafic rocks (diabase), their parent materials and other factors resulting in the formation of eutrophic soils having A chernozemic horizon associated with Red Nitosols (Alfisol) and Red Latosols (Oxisol) of the Amazonian environment. In seventh chapter, the flexural behavior of laterally loaded tapered piles in cohesive soils is investigated. Eighth chapter shows that for the same load cases, the values of the torsion moment and shear stress are not significant those of bending moments and normal stresses, respectively. In ninth chapter, the mineral and chemical composition and several essential physical parameters of unheated expansive soil are indicated by XRD and EDX analysis. Tenth chapter provides a quantitative framework for estimating energy requirements for soil penetration work done by earthworms and plant roots, and delineates intrinsic and external mechanical limits for bioturbation processes. Last chapter presents characteristics of soil fractal features to different land uses in typical purple soil watershed.

Chapter 1

CHEMICAL AND HYDRAULIC BEHAVIOR OF A TROPICAL SOIL COMPACTED SUBMITTED TO THE FLOW OF GASOLINE HYDROCARBONS

Wisley Moreira Farias[1], Geraldo Resende Boaventura[1], Éder de Souza Martins[2], Fabrício Bueno da Fonseca Cardoso[3], José Camapum de Carvalho[1], and Edi Mendes Guimarães[1]

[1]Universidade de Brasília, Brasília-DF, Brazil

[2]Embrapa/ Cerrados, Planaltina –DF, Brazil

[3]ANA-Agência Nacional de Águas –Brasília-DF, Brazil

INTRODUCTION

Gasoline is a fuel comprised basically of hydrocarbons such as aromatic, olefinic and saturated compounds of a carbon chain comprised of 4 to 12 atoms. The aromatic compounds such as benzene, toluene, ethylbenzene, o-, m-, p-xylene (BTEX) are harmful to human health (Cairney et. al., 2002). As these compounds are harmful to health, the legislation becomes restrictive. The U.S. Environmental Protection Agency for drinking water (US EPA) establishes the maximum concentration of benzene in $5\mu g.L^{-1}$. In Brazil the Ordinance of the Ministry of health number 2,914 in 12th December 2011, stipulates that the maximum allowable concentration of benzene is 5 $\mu g.L^{-1}$ regulation of drinking water contaminant. Soil in contaminated residential areas, Brazil has been adopting as intervention guide value, the concentration of benzene 0.08 $mg.kg^{-1}$ set by the State of São Paulo in 2001. This value indicates the intervention limit of contamination where there is potential risk to human health.

Brazil produces type-C gasoline which is different than other types due to its anhydrous alcohol content (ethanol), in the proportion of 25% (Farias, 2003). The alcohols are soluble in water, and have a significant mobility potential to percolate through the soil until reaching underground water (Ulrich,

1999; Corseuil and Fernandes, 1999). The alcohol in gasoline in an aqueous medium promotes co-solvency which is the increase in the solubility of the hydrocarbons in the gasoline in an aqueous solution (Banerjee and Yalkowsky, 1988; Cline *et al.*, 1991).

Solubility is generally controlled by the polarity effect, which decreases in size for molecules with the same organic function. Non-polar or weakly polar substances dissolve in similar solvents. Thus, highly polar compounds dissolve in polar solvents such as water. The polarity or dipolar moment is proportional to the dielectric constant, and therefore high dielectric constant compounds (values of 80 for water and 34 for methanol) dissolve ions through hydration of the disassociated types (Fernandez and Quigley, 1985).

On the surface of clay-minerals, the absorbed water forms a double layer, which reduces the strength of interaction between the negatively charged clay particles and the cations in the colloidal solution. The hydrophobic hydrocarbons in the gasoline have low dielectric constant values, thus provoking the collapse of the double layer. This collapse is due to the contraction of the double layer through the attraction of the contra-ions which are closer to the superficial charge of the clay-minerals, favoring flocculation, and consequently the increase in permeability due to the increase in pore space (Mesri and Olson, 1971; Fernandez and Quigley, 1985 and 1988).

The co-solvêncy is responsible for the partition of BTXs to the aqueous phase, promoting the reduction of density of colloidal solution of soil, providing increased viscosity and a reduction of the surface tension (Mcdowell and Powers, 2003). This reduction in surface tension and generated by the collapse of the electrical double layer that there was between the soil and water (Farias, 2003).

Aspects of the Transport of Pollutants in Soils

The transport of pollutants in the soil can occur through the porous medium and saturated or unsaturated fractured media. This transportation occurs through physical or chemical processes, or through both processes. The chemical process becomes evident when the velocity of the fluid is not sufficiently high (i.e., less than 10^{-6} cm/s), generating a gradient due to the flow of the solute (contaminating agent) from the more concentrated medium to the less concentrated one. This process is called molecular diffusion (Rowe, 1988; Pastore and Mioto, 2000). This type of flow has been widely studied with metals and organic compounds in solid waste landfill leachate contaminants, for application in compacted soil layer, also called liners (Shackelford and Daniel, 1991; Rowe, 1988; Barone *et al.*, 1988).

Fernandez and Quigley (1985) developed an experimental research program to evaluate the hydraulic behavior of clayey-like soil (Sarnia, Ontario), permeated with liquid substances such as benzene, xylene, cyclohexane, aniline, propanol, acetone, alcohol and water. The results have shown that Hydraulic conductivity increased from 5×10^{-9} to 1×10^{-4} cm.s^{-1} along with a decrease in the dielectric constant from 80 (water) to 2 (benzene).

When there is a hydraulic gradient, the velocity of the solvent is relatively high and the transportation of the solute is practically managed by the velocity of the solvent, a mechanism which is known as an advection process. In this process, the velocity of the fluid is governed by Darcy's Law, which considers not only the characteristics of the soil, but also those of the fluid (Fernandez and Quigley, 1988).

In order to have good performance, the compacted clay liners must have a hydraulic conductivity less than 10^{-8} cm/s. However Daniel and Koerner (1995) defined that the hydraulic conductivity of clay liners must be less than or equal to 10^{-7} cm/s. This low flow is normally associated with the presence of clay-minerals, and at least 15 to 20% of particles with sizes under 2 mm, as well as a minimum plasticity greater than 7%, activity greater than 0.3, and cation exchange capacity (CEC) greater than 100 mmol$_c$/dm^3 of soil (Rowe *et al.*,1995).

The natural organic material of the soils have proven to be efficient in the retarding process through the sorption of hydrophobic hydrocarbons, which are also found in gasoline (Chiou *et al.*, 1983; Karickhoff*et al.*, 1979; Schwarzenbach *et al.*, 1993).

Importance of Research

The aim of this study is to evaluate the behavior of a tropical soil, and their performance as liner against the flow of hydrocarbons from gasoline, by interpreting transportation according to physical and chemical parameters, as well as micromorphological aspects. For this characterized the mineralogy of the soil and the influence of his organic matter (OM), considering the adsorption processes of hydrocarbons from gasoline and hydraulic behavior in the laboratory by variation of the hydraulic gradient in front of the gasoline flow through compacted soil. This study also aims to contribute to the understanding of the dynamics of the flow through the soil of specific groups of compounds: aromatic, olephine, saturated hydrocarbons and the ethanol found in Brazilian type-C gasoline (a complex mixture of organic compounds).

LOCATION AND SOIL CLASSIFICATION

The soil sample was collected indisturbed block in depth of 4 m in the experimental field of foundations and test field of the Civil Engineering Department of the University of Brasília, located on the University campus in the city of Brasília, Brazil with coordinates 15° 56 ' 45 "S, 47° 52 ' 20" W (Fig. 1).

The sample of lateritic soil typical of the Brazilian *cerrado* region was studied. According to the Brazilian Soil Classification System (Embrapa, 1999), the soil was classified as Red Latossoil, considered as Ustic Rhodic Oxisol according to the U.S. Soil Taxonomy and Geric Ferralsol Ferric (FAO, IUSS Working Group WRB, 2007). It possesses a silt-clay-like texture, a large quantity of granular aggregates, and small pores. Visually, it is homogeneous and isotropic, without the presence of discontinuities.

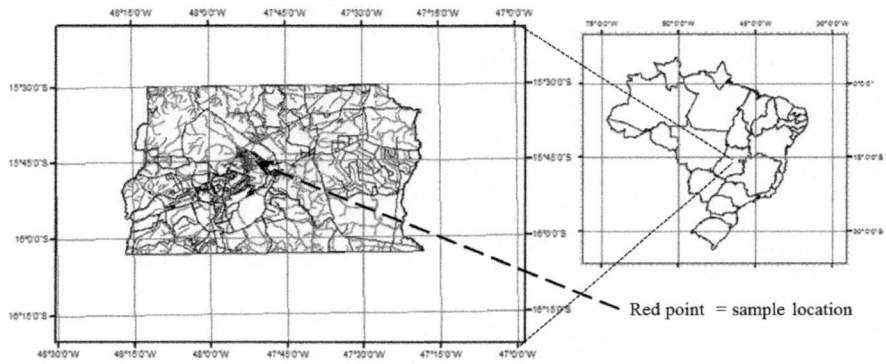

Figure 1. Map of location of the soil collection.

METHODOLOGY

The characterization of the soil involved the use of physical, chemical and mineralogical analysis.

Physical Tests

Geotechnical tests of physical properties of soils were performed following the Brazilian Association of Technical Standards (ABNT) procedures: test of limits of consistency called Atterberg limits following the ABNT NBR 7180/84 plastic limit; 6459/84 liquid limit following the Casagrande method. Before the grain-size determination, the real density was determined according the ABNT NBR 6508/84 method. The grain-size distribution curve was determined using a grain-size digital meter Malvern Mastersizer with lens de 300Rf for grain size range of 0.05 μm to 900μm at 25 ° C. For this analysis, the sample was previously passed through a No. 40 sieve. The analyses of the samples were done either with or without ultrasound dispersion. Ultrasonic condition was 5 minutes of dispersion in distilled water with ultrasonic level set at 5. The grain size fractions were classified following the Brazilian standard NBR 6502/93.

The degree of flocculation and dispersion of soil particles was determined comparing the results of grain size determinations before and after ultrasonic dispersion.

Hydraulic Conductivity

The test of hydraulic conductivity in compacted soil in standard Proctor energy were performed in a conventional manner with water using the variable charge and special form for gasoline (Fig. 2 and Fig. 3).

Figure 2. Hydraulic Conductivity cell.

The gasoline hydrocarbons, for possessing volatile and low-density compounds require a special sealed cell to avoid losses due to evaporation and leakage and to support the applied tensions. The material selected for the construction of the special cells was stainless steel, to avoid reaction and adsorption problems in the walls, which is the case of plastics and acrylics (Doanhue *et al.*, 1999).

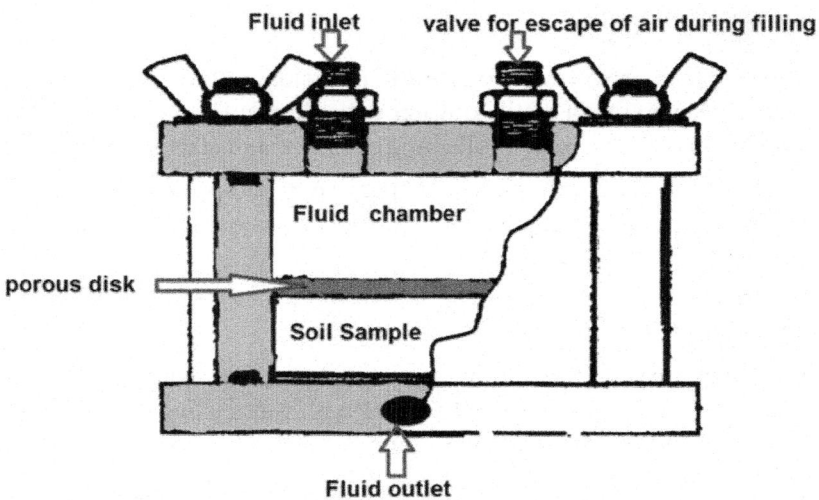

Figure 3. Schematic of permeameter Cell.

The system used to perform the gasoline's hydraulic conductivity was similar to that applied by Fernandes (1989). The special cell may be disassembled, and is made up of three parts. The first part is a cylinder, where the test material and reservoir are found. This part is 5 mm thick, 110 mm long and has an internal diameter of 77.2 mm. The other two parts are the upper and lower lids. Both have cavities filled with rubber rings which are able to prevent the reaction of the hydrocarbons in the gasoline and act as a seal when the cylinder is assembled. The upper lid has two openings, one for the entry of fluid and the other for the application of vertical tension with compressed air. The lower lid is made up of an outgoing flow register which is connected to a collecting container. The two lids are 120 x 120 mm² square, 10 mm thick. The connections were made out of aluminum, due to its low cost and flexibility; the connecting joints were sealed with 3M automotive glue and winding sealing thread, in order to prevent leaks and to make the system more secure.

The conductivity test was performed with test material 5 cm long, compacted at normal Proctor energy at optimal water content condition, in the cylinder of the hydraulic conductivity cell. Then, a thin disk of porous ceramic was placed top of the sample. The small space between the disk and the cylinder wall was filled with 3M glue to prevent preferential flows along the wall, and to ensure that the gasoline only passed through the porous ceramic disk. The cell was then assembled, and the upper and lower lids were connected to the cylinder. The cylinder is 11 cm high, of which the remaining 6 cm were filled with type C gasoline. After the cell was totally sealed and connected to the compressed air system, with pressure controlled by a manometer, it was connected with plastic tubes able to support high pressure. The conductivity tests were performed for various applied vertical pressures. For each pressure applied, the hydraulic conductivity was measured. The pressures were varied to see how the soil sample behaved with an increase in hydraulic gradient upon the flow of gasoline. The hydraulic conductivity was measured in the laboratory at static tensions σ_v of 50, 100, 150, 200, and 300 kPa, with respective hydraulic gradients of 75, 150, 225, 300, and 450.

The residual water of the soil pores mixed with gasoline collected in the test was previously run through a separating funnel to remove the aqueous phase to later take a reading of hydrocarbons of the gasoline through infrared technique.

The test material of the lateritic soil sample, before and after the hydraulic conductivity test conducted with water, and the other with the flow of gasoline, were dried at room temperature. Micromorphological analyses were performed on Thin Lamina (TL) in vertical sections, prepared by impregnating the sample with plastic resin (Cardoso, 1995; Martins, 2002). The instrumental technique used for the microscopic views of the TL was Optical Microscopy.

Mineralogical Characterization

The identification and quantification of minerals in the sample were carried out by the method developed by Martins (2000). This method involves the use of X-ray diffraction (XRD) technique for identifying the minerals, chemical analysis for the determination of major elements (Al, Fe, Si, Ca, Mg and Ti), thermogravimetric analysis (TGA), and the use of Munsell color code (Munsell color company Inc., 1975). The determination of major chemical elements was performed by ICP-AES after the fusion of samples with alkaline NaOH as fondant at a temperature of 450 ° C for 40 minutes using the nickel crucible. Determinations of elements by ICP / AES (atomic emission spectrometry of Plasma Induced Coupling) were performed with Thermo Jarrell ASH equipment, model Iris / AP.

The thermogravimetric analysis were applied to quantify the kaolinite and gibbsite. For this used the TGA Shimadzu equipment with temperature ramp of 20 °C to 1500 °C, with speeds ranging from 0.2 to 60 ° C / min, using the software TAS60WS for the treatment of data. The Munsell code was used for determining the ratio of hematite and goethite in the soil samples. The CEC of soil was determined using the principle of the simple as the sum total of the exchangeable cations that a soil can adsorb. The determination of the organic matter content was done prior to extraction using wet oxidation method with potassium dichromate in sulfuric medium. The excess of dichromate after oxidation was titrated with standard solution of ferrous ammonium sulfate (Mohr salt).

Chemical Characterization

The pH was measured in the soils samples in distilled water medium using a combined glass electrode Ag/AgCl (potentiometric method).

In order to study the influence of OM and mineralogy in the gasoline sorption process, an experiment was performed with samples treated with H_2O_2 and another without treatment.

The extraction of the OM used 15 g of soil in a porcelain capsule, with 10 mL of H_2O_2 volume 30% and with agitation in a 50 mL Becker cup. After agitation, there was an effervescent reaction, when the capsule was covered with clock glass for one night. The process was repeated until the complete disappearance of the reaction. It was then washed 3 to 5 times in distilled water, using a Büchner funnel with filtering under reduced pressure. Then, for the gasoline sorption test, the sample was allowed to dry at room temperature.

The sorption test procedure used 2 g of soil with 25 mL of gasoline placed in an amber glass jar under agitation for 24 hours at a temperature of 22°C. After this, the samples were centrifuged as in the processes described above, with the removal of a 15 mL portion for analysis.

The hydrocarbons content of the gasoline samples was determined at the National Petroleum Agency (ANP) laboratory, in Brasilia, with a (FTIR = Fourier Transform Infrared), manufactured by Grabner Instruments, model IROX 2000. This instrument qualified and quantified the compounds, generating the mass and volume percentages of the ethanol, aromatic, olephine and saturated compounds.

RESULTS

Tab. 1 and 2 present data of the physical, chemical and mineralogical Brazilian soil and constituents of the gasoline type C studied.

Table 1. Characteristics of the soil (Farias, 2003).

Test	Lateritic
Atterberg Limits	
Liquid limit-W_L (%)	41
Plastic limit-W_p (%)	29
Plastic Index-I_p (%)	12

Activity	0,18
Grain size distribution*	
Clay (%)	65
Silt (%)	34
Sand(%)	1
Degree of flocculation (%)	92
Degree of dispersion (%)	8
Chemical Parameters	
pH	5,70
Organic Matter content (%)	0,41
CEC ($mmol_c/dm^3$)	6,4
Mineralogy	
Quartz (%)	30,2
Anatase (%)	1,57
Kaolinite(%)	24,6
Gibbsite (%)	25,5
Goethite (%)	4,6
Hematite (%)	7,5
Illite (%)	2,2
Vermiculite (%)	3,7
Hydraulic Conductivity in water (cm/s)	3,7.E-07

[i] - *Grain size data obtained by ultra-sound waves using a laser beam grain size analyser.

Tab. 2 presents the composition of the Brazilian type-C gasoline, according to Farias (2003).

Table 2. Brazilian Type C gasoline data.

Compounds	**Mass (%)**
Aromatics	20,8
Olefins	22,4
Saturated	31,4
Ethanol	25,4

Fig. 4 presents the increase in hydraulic conductivity with an increase in the hydraulic gradient. At a gradient of approximately 210, conductivity becomes practically constant. Fig. 5 presents the intrinsic permeability, which considers the characteristics of the soil, but does not consider the chemical and physical properties of the fluid. Intrinsic permeability reaches values close to $10^{-13}m^2$. However, as the hydraulic gradient increases, stability reaches approximately $10^{-11}m^2$.

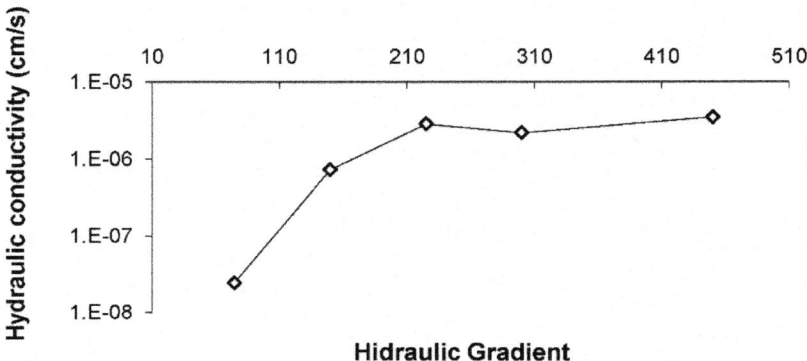

Figure 4. Behavior of hydraulic conductivity and hydraulic gradient of laterite soil on the gasoline flow.

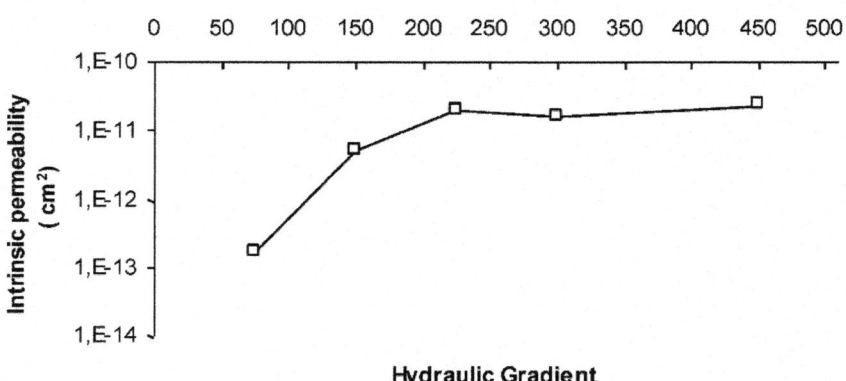

Figure 5. Behavior of the intrinsic permeability and hydraulic gradient of laterite soil on the gasoline flow.

Fig. 6 depicts the behavior of the hydraulic conductivity relative to the volume of pores while undergoing saturation in the test material with gasoline at a tension of σ_v of 50 kPa. The saturation process takes place with the expulsion of the interstitial water accumulated in the pores due to optimal compacting moisture content (w_{opt} = 26%) is the test material at normal Proctor energy. It may be observed that as the volume of pores in the gasoline flow increases, conductivity decreases from 4 to 2 x 10^{-8} cm.s^{-1}. This suggests that the behavior of the reduction may be represented by a second-order equation.

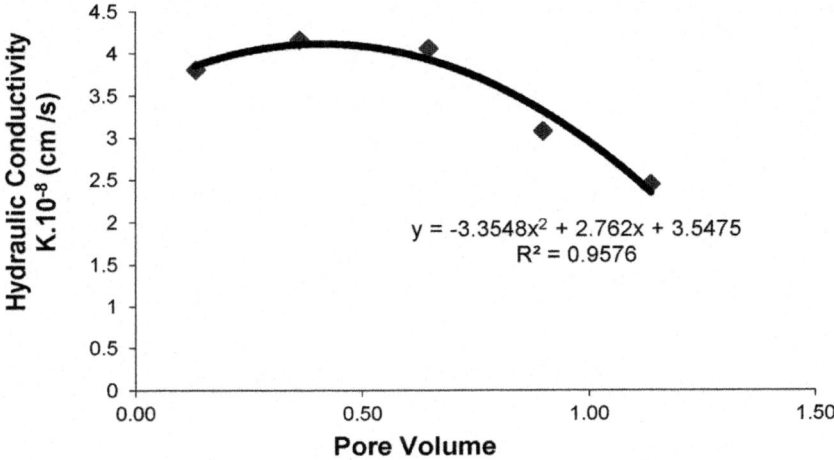

$$y = -3.3548x^2 + 2.762x + 3.5475$$
$$R^2 = 0.9576$$

Figure 6. Behavior of the lateritic soil saturated with gasoline at 50 kPa.

Fig. 7 presents the saturation process at a σ_v tension of 50 kPa, based on the ratio between the concentration (C) of the gasoline hydrocarbons passing through the soil sample, and the initial concentration (C_o) added to the reservoir, in relative to the volume of pores. The hydrocarbons concentration data are from the Light Non-aqueous Liquid Phase (LNALP), after the flow through the soil sample in the hydraulic conductivity test.

Table 3. Result of the physical parameters of the test material.

Sample	w (%)	γ (kN.m^{-3})	γ_{dmax} (kN.m^{-3})	γ_s (kN.m^{-3})	e	n	S r (%)	Vv cm3
lateritic*	1,7	17,7	17,4	27,5	0,58	0,4	8,1	134,3
lateritic**	1,7	15,8	15,6	27,5	0,77	0,4	6,2	178,5
lateritic***	1,8	14,7	14,5	27,5	0,90	0,5	5,3	210,0

[i] - *Dry soil sample before the hydraulic conductivity test

[ii] - **Dry soil sample after the hydraulic conductivity test with the water flow

[iii] - *** Dry Soil sample after the hydraulic conductivity test with the gasoline flow

The results in Tab. 3 present the physical parameters of the compacted test materials dried at room temperature before and after the hydraulic conductivity test. Highlights the volume of voids (**Vv**), which changes substantially when there is a flow of gasoline. The degree of saturation also decreases after the flow of gasoline.

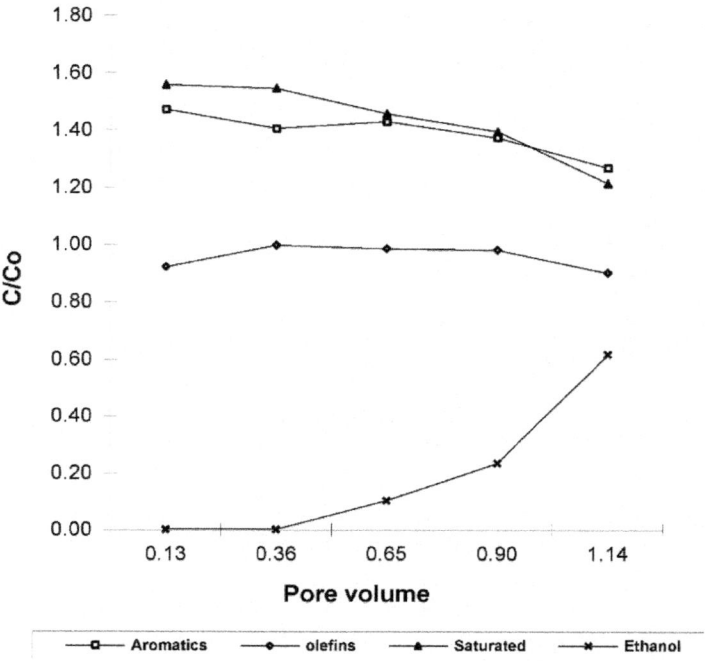

Figure 7. Light non-aqueous liquid phase ratio of the gasoline relative to the volume of pores of the lateritic soil in a saturation process at 50 kPa.

The micromorphology of the three compacted soil samples was important in order to visualize the behavior of the test material before the hydraulic flow (Fig. 8), after the hydraulic flow with water, and after the flow with gasoline. It must be noted that the grains of quartz make up approximately 40% of the total solid material; variable in size, 0.12 mm on average; and overall, are sub-rounded to angular. They are highly fractured, without orientation and their contours present corrosion. In spite of the compacting, the structure of this soil is not totally dispersed, for microaggregations of oxyhydroxides of Fe and Al remain, forming micropores. The compacted soil sample submitted to percolation in water showed a single micro-structural difference relative to the one performed on the LT of the compacted soil sample. Actually, there was an increase in small canal-type voids, generated by the flow of water (Fig. 9). The micromorphology regarding the LT of the compacted soil submitted to the flow of gasoline also showed only a quantitative increase in canal-type voids (Fig. 10). However, this variation was greater than that registered in the previous sample with the water flow.

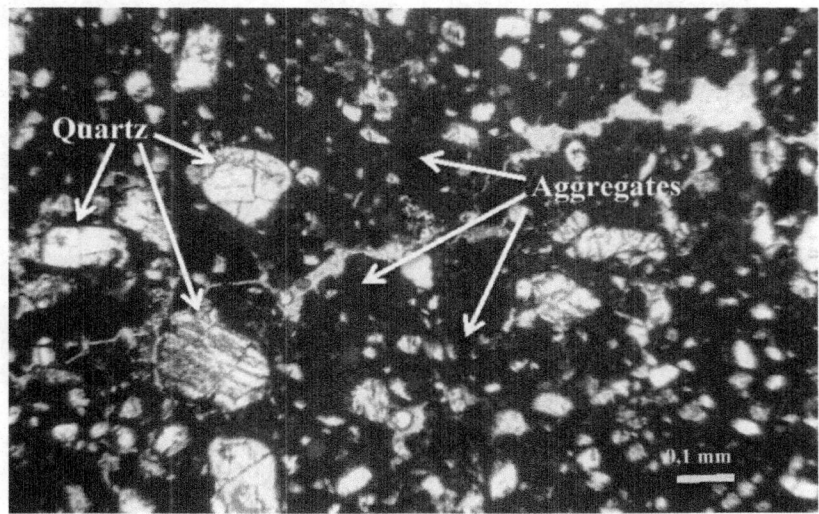

Figure 8. Photomicrography of the porfirosquelic APE, aggregates, and quartz grains of the compacted lateritic soil. Parallel nichols (N//).

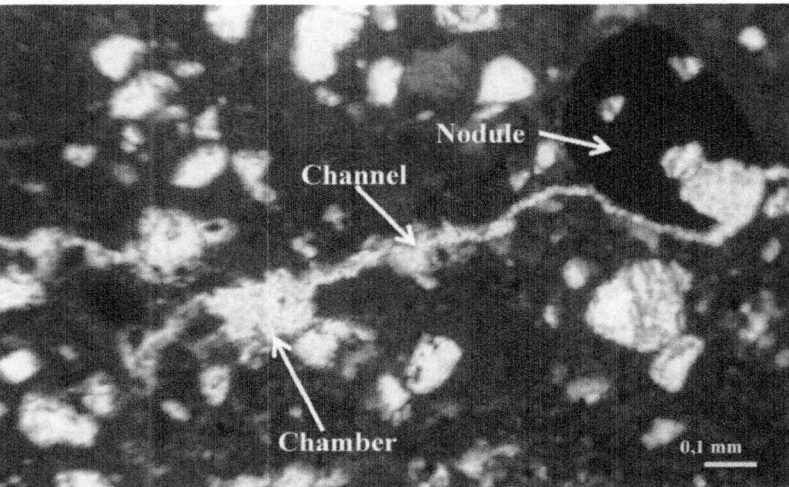

Figure 9. Photomicrography showing the nodules and canal- and chamber-type voids of the compacted lateritic soil submitted to percolation with water. Parallel nichols (N//).

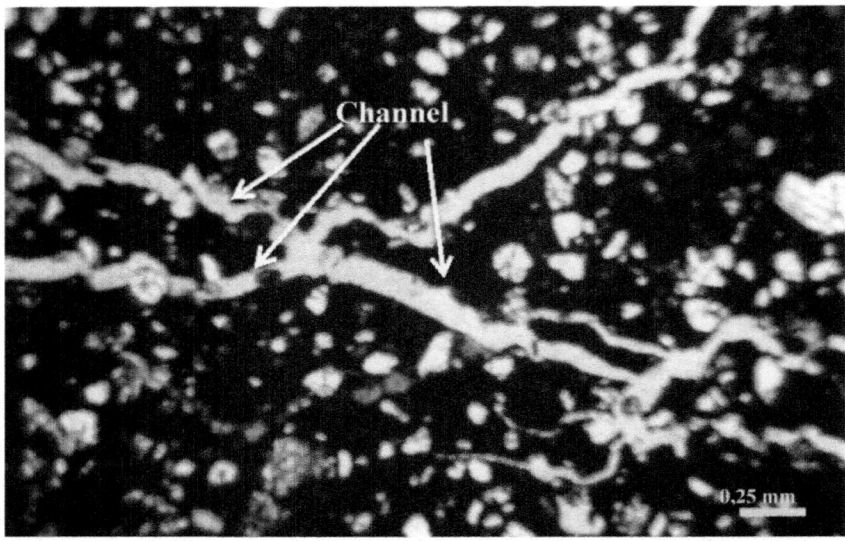

Figure 10. Photomicrography showing the canal-type voids of the compacted lateritic soil submitted to percolation with gasoline. Parallel nichols (N//).

Fig. 11 presents the results of the adsorption of the ethanol and aromatic substances in the samples with and without the extraction of organic matter with the use of hydrogen peroxide. Note that the samples treated with extractor presented low adsorption. Aromatic compounds showed no adsorption after extraction of organic matter contained in the soil.

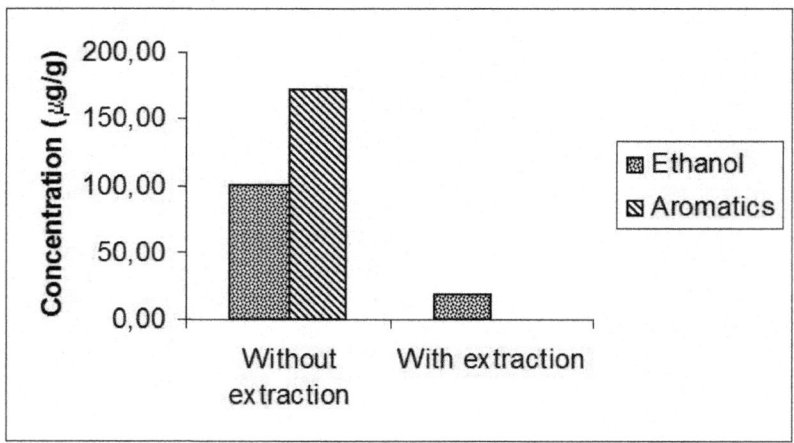

Figure 11. Results of the adsorption of the gasoline hydrocarbons in the soils with and without the extraction of the soil organic matter.

Gasoline ethanol can be adsorbed on the sites of hydroxyls of the octahedron of Al, exposed by fractures, Scrubs or crystalline lattice imperfections, or by interactions with the Fe oxides and hydroxides and Al amorphous. This occurs from adsorption of hydrogen bonds, which can also occur with water strongly adsorbed on the surface of the clay minerals (Fig. 12).

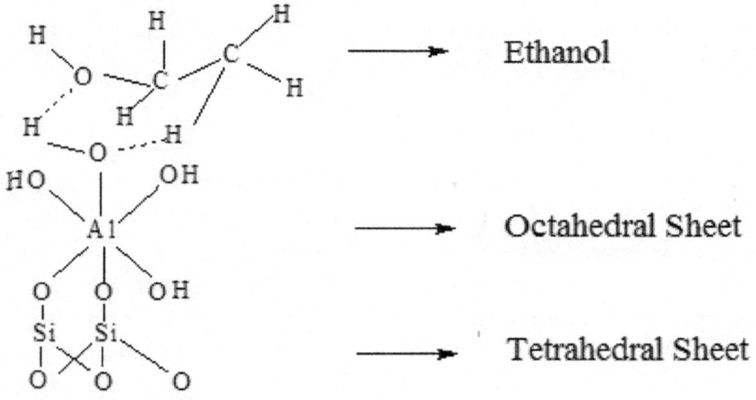

Figure 12. Coordination of interaction of hydrogen and hydroxyl ethanol exposed in the clay mineral (1:1).

DISCUSSION OF THE RESULTS

The discussion of the results is focused on three main aspects. The first considers mineralogical, chemical and physical characteristics of the material with potential for liners. The second aspect is assessed the performance of Laterite soil on gasoline hydrocarbon flow subjected to high hydraulic gradients, causing an acceleration of the process of formation of flow channels for compressed soil in power of Proctor. The last aspect to be evaluated is the power of gasoline hydrocarbon adsorption by soil with OM and no OM.

The Delimiters Criteria Material with Potential For Liners

Evaluating the criteria prescribed by Rowe et al., (1995) the soil presents considerable levels of Fe oxides and hydroxides and Al (hematite, goethite and gibbsite) and kaolinite with only 30.2% of quartz. As the mineral vermiculite is low soil activity levels was 0.18, less than the 0.3 suggested by the literature. However this value of activity indicates that the material is not expandable, being a good quality for liners. The cationic exchange capacity (CEC) also presented low value (6.4 $mmol_c/dm^3$) comparing with value defined in literature. Tropical soils lateritic in general are highly weathered with low or no mineral content of 2:1, which are typical of temperate climate. Therefore, the activity and CEC are low. The granulometry performed 65% clay fraction indicated more than 20% of particles less than 2 mm confirming the material rich in clay fraction indicating low permeability material when compressed. Thus, the hydraulic conductivity parameter value introduced into water in the order 10^{-7}cm/sec subjected to a pressure of 20 kPa (Tab. 1). These results of the hydraulic conductivity characterize the material with great potential for liner according to predefined values in the literature.

Lateritic Soil Performance as Liner

The hydraulic conductivity of gasoline type C brazilian obtained values between 10^{-8} to 10^{-7} cm/s to a gradient of 75 with a pressure of 50 kPa, which corresponds to 5 m of the water column (Fig. 4). Such a result of hydraulic conductivity defines the material as great for barrier on gasoline hydrocarbon flow according to predefined values in literature (Rowe et al., 1995; Daniel and Koerner, 1995). With the increase of the hydraulic gradient there was an increase in hydraulic conductivity until it reaches a level of stabilization in gradient greater than 210 (Fig. 4). Although it does not occur to the destruction of the liner to the 210 to 450 gradients suggests avoid gradients greater than 100 in the projects of protection of underground fuel tanks, ensuring in this way a hydraulic conductivity around 10^{-7} cm/s for liners according to the literature. The intrinsic permeability or specific considers simply the porous medium, not considering the characteristics of fluid. The values found for intrinsic permeability compacted laterite soil is similar to that found in the literature to clay (Freeze and Cherry, 1979).

The compacted soil voids indexes suffered increased 0.58 before tests to 0.77 with water flow and 0.90 with gasoline flow in hydraulic gradient of 75. The empty volume also increased from 134.3 to 210.0 (Tab. 3). The soils studied presented a high degree of flocculation due to the aggregates of the oxyhydroxides of Fe and Al. Even when compacted, they contain micro-aggregates which are not destroyed. When a flow is established through the soil, the micro-aggregates may interconnect, forming flow channels. The physical behavior provoked by the flow may be visualized in the micromorphology of the samples in Fig. 8, 9 and 10. However even with these micros channels formed in the compacted soil hydraulic conductivity limit of 10^{-7} cm/s (Daniel and Koerner, 1995) is not affected considering a gradient of 100.

Adsorption Performance for Hydrocarbons of Gasoline

The performance of laterite soil to a gasoline hydrocarbon flow subjected to a pressure of 50 kPa with 75 gradient was evaluated for pore volume and the ratio C/Co in the process of saturation of the compacted clay liner for gasoline, in other words, there was the expulsion of the water contained in the soil by the process of compression to achieve the optimum water content of compaction. The reason indicates that values above 1 there is an LNAPL phase concentration of groups of substances evaluated. The groups were evaluated for aromatic, olefins, saturated and ethanol.

In Fig. 7, the aromatic compounds appear as constants in the saturation process. Since they are hydrophobic, their polarity is low, and are more easily transported in the soil. The olefines and saturates have a greater C/Co ratio in the LNAPL due to their low solubility in water, being lower than the aromatic compounds, which are more affected by ethanol through co-solvency. In the 0.13 a 0.36 pore volume range, the ethanol is partitioned to the aqueous phase and, as the saturation of the pores with gasoline increases, the C/Co ratio for ethanol in the LNAPL also increases. The partitioning of the ethanol for the aqueous phase is natural and is due its polarity, which makes it mixable in water. Thus, the ethanol, along with the other hydrophobic compounds in the gasoline, favors the collapse of the double layer, as well as the increase in micropores (Rowe et al., 1995)

The results in Fig. 11 show that the soil organic matter, although in low quantities, has an influence of almost 0.41 %, in the sorption process. The soil studied was collected at a depth of 4 meters, thus contained evolved organic matter, possibly fulvic acid. The removal of the organic material with hydrogen peroxide showed low ethanol sorption. The aromatic compounds, which are hydrophobic, were not absorbed.

Evaluating the transportation of gasoline compounds by soil with application of 50 kPa of pressure, indicates low retention and greater mobility, because mostly they are hydrophobic compounds that do not bind the soil particles. Another aspect of this experiment is that has not been evaluated by diffusion flux, which occurs at speeds equal to or less than 10^{-10} cm/s. The test of sorption to organic matter proved to be important in the retention process of ethanol. In view of the low adsorption of gasoline compounds by soil suggests considering projects of liners gradients below 75 and pressures less than 50 kPa ensuring a hydraulic conductivity greater than 10^{-8} cm/s and use clayey material rich in organic matter to promote greater retention of ethanol and avoid or reduce the effect of co-solvency.

CONCLUSION

Since the lateritic soil studied possesses a high aggregation capacity, even when compacted at normal Proctor energy, micropores remain which, in high hydraulic gradient situations, are interconnected, forming flow channels.

However, even under higher hydraulic gradients in the gasoline percolation tests, this soil presents good material of liners. This is due to the stabilizing of the flow channels formed, favoring also the stabilizing hydraulic conductivity.

The measure that gasoline occupies the pores in the process of saturation the concentration of ethanol increases. This is due to the polarity of the ethanol. The aromatic compounds maintain a C/C_0 ration close to 1, as the volume of pores increases, indicating that these are tracers due to their low dielectric constant and polarity. Due to their low solubility in water, the olephines and saturates are more present in the LNAPL phase. These hydrocarbons may form emulsions, favoring transportation through the soil in the aqueous phase.

Regarding the retarding potential of the lateritic soil, evaluated by the sorption parameter, it is not directly correlated with the mineralogy, because the aromatic compounds are not absorbed when the organic material is extracted. Actually, this sorption may be correlated with a certain type of humic substance, which may be interacting with the poly-amorphs of the oxyhydroxides of Fe and Al in the soil, favoring interaction with the aromatic compounds.

Finally, a low hydraulic gradient context (< 75), hydraulic conductivity $<$ 10^{-8} and organic matter, in lateritic soil can improve the performance of liner.

ACKNOWLEDGEMENTS

The authors are gratefully to the Conselho Nacional de Desenvolvimento Científico e Tecnológico – CNPq, CAPES and ANP for the fellowships and financial support granted to the accomplishment of this research.

REFERENCES

1.   ABNT (1984) Solo – Determinação do limite de liquidez –NBR 6459/84. Associação Brasileira de Normas Técnicas, Rio de Janeiro, RJ, 6p.

2.   ABNT (1984) Solo – Determinação do limite de plasticidade – NBR7180/84. Associação Brasileira de Normas Técnicas, Rio de Janeiro, RJ, 3p.

3.   ABNT (1984) Solo – Grãos de solo que passam na peneira de 4,8 mm - Determinação da massa específica (método de ensaio) – NBR 6508/84. Associação Brasileira de Normas Técnicas, Rio de Janeiro, RJ, 8p.

4.   ABNT (1993) Rochas e solos – Terminologia – NBR 6502/93. Associação Brasileira de Normas Técnicas, Rio de Janeiro, RJ, 19p.

5.   Banerjee, S. and Yalkowsky, S.H. (1988). Cosolvent-Induced Solubilization of Hidrophobic Compounds into Water. Analytical Chemistry, v.60, p. 2153-2155.

6.   Barone, F.; Yanful, E.K.; Quigley, R.M. and Rowe, R.K. (1988). Effect of multiple contaminant migration of diffusion and adsorption of some domestic waste contaminants in a natural clayey soil. Geotechnical Research Report GEOT -5-88, Geotechnical Research Centre. The University of Westem Ontario, London, Ont.

7.   Cairney, S., Maruff, P., Burns C., Currie B (2002) The neurobehavioural consequences of petrol (gasoline) sniffing. Neuroscience and Biobehavioral Reviews, v.26:1, p. 81-89.

8.   Cardoso, F. B.F (1995) Análise Química, Mineralógica e Micromorfológica de Solos Tropicais Colapsíveis e o Estudo da Dinâmica do Colapso. Dissertação de Mestrado, Departamento de Engenharia Civil, Universidade de Brasília, Brasília, DF, 142p.

9.   Chiou, C.T., Porter, P.E. and Schmedding, D.W. (1983). Partition equilibrium of nonionic organic compounds between soil organic matter and water. Environmental Science and Technology, 17(4):227-231.

10.   Cline, P.V., Delfino, J.J. and Rao, P.S.C. (1991). Partitioning of aromatic constituents into water from gasoline and other complex solvent mixtures. Environmental Science and Tecnology, 25(5):914-920.

11.   Corseuil, H. X. ; Fernandes, M.(1999). Efeito do Etanol no Aumento da Solubilização de Compostos Aromáticos Presentes na Gasolina Brasileira. Revista Engenharia Sanitária e Ambiental, v. 4, n. 1 e 2, p. 71-75.

12.   Daniel, D.E and Koerner, R.M. (1995) Waste containment facilities – Guidance for construction, Quality Assurance and Quality Control of Liner and Cover Systems, American Society of Civil Engineers, ASCE Press, New York, 354 pp.

13.   Donahue, R.B., Borbour, S.L. and Headley, J.V. (1999). Diffusion and adsorption of benzene in Regina clay. Canadian Geotechnical Journal, 36(3):430-442.

14.   FAO, IUSS Working Group WRB (2007). World Reference Base for Soil Resources 2006, first update 2007. World Soil Resources Reports No. 103, Rome.

15.   Farias, W.M. (2003). Condutividade Hidráulica de Solos Tropicais Compactados a Hidrocarbonetos da Gasolina. M.Sc. thesis, Department of Civil and Environmental Engineering, University of Brasilia, Brasilia DF, 152 pp.

16.   Fernandez, F. and Quigley, R.M. (1985). Hydraulic conductivity of natural clays permeated with simple liquid hydrocarbons. Canadian Geotechnical Journal, 22:205-214.

17. Fernandez, F. and Quigley, R.M. (1988). Viscosity and dieletric constant controls on the hydraulic conductivity of clayey soils permeated with water – soluble organics. Canadian Geotechnical Journal, 25:582-589.

18. Freeze, R. A. and Cherry, J. A (1979). Groundwater. Prentice-Hall, New Jersey, USA, 604p.

19. Karickhoff, S.W., Brown, D.S. and Scott, T.A. (1979). Sorption of hydrophobic pollutants on natural sediments. Water Research, 13:241-248.

20. Martins, E.S., (2000). Petrografia, Mineralogia e Geomorfologia de Rególitos latériticos do Distrito Federal. Tese de Doutorado, Instituto de Geociências, Universidade deBrasília, Brasília, DF, 196 p.

21. McDowell, C.J. and Powers, S.E. (2003). Mechanisms Affecting the infiltration and distribution of ethanol-blended gasoline in the vadose zone. Environmental Science and Technology, 37:1803-1810.

22. Mesry, G. and Olson, R.E. (1971). Mechanisms controlling the permeability of clays. Clay and Clay Minerals, 19(3):151-158.

23. Mussell Soil Color Company (1975) Mussell soil color charts. Baltimore, 1V, 117p.

24. Pastore, E.L. and Mioto, J.A. (2000). Impactos ambientais em mineração com ênfase à drenagem mineira ácida e transporte de contaminantes. Solos e Rochas. São Paulo, SP, 23(1):33-53.

25. Rowe, R.K. (1988). Contaminant migration though groundwater: The role of modeling in the design of barriers. Canadian Geotechnical Journal, 25(4):778-798.

26. Rowe, R.K., Quigley, R.M. and Booker, J.R. (1995). Clayey Barrier Systems for Waste Disposal Facilities. E and FN Spon, London, England, 390p.

27. Scharzenbach, R.P., Gschwnd, P.M. and Imboden, D.M. (1993). Environmental Organic Chemistry. Wiley Interscience, New York, NY, 1313p.

28. Shackelford, C.D. and Daniel, D.E. (1991). Diffusion in saturated soil. II: Results for compacted clay. Journal of Geotechnical Engineering, 117(3): 485-506.

29. Ulrich, G. (1999). The Fate and Transport of Ethanol-Blended Gasoline in the Environment. Governors' Ethanol Coalition, Lincoln, Nebraska, 88p.

Chapter 2

DETERMINATION OF HYDRAULIC CONDUCTIVITY BASED ON (SOIL) - MOISTURE CONTENT OF FINE GRAINED SOILS

Rainer Schuhmann, Franz Königer, Katja Emmerich, Eduard Stefanescu, and Markus Stacheder

Karlsruhe Institute of Technology (KIT) Competence Centre for Material Moisture (CMM) Karlsruhe, Germany

INTRODUCTION

The chapter will be divided into the subchapter material, processes and systems. The first one will focus on the physical, chemical and dynamic material properties and their measuring methods. The second specifies the dynamic of the surface moistening and fluid flow. The comprehensive characterization of materials is prerequisite to understand processes in large (geo)-technical systems and their manipulation. The transition from nano (material) via meso (processes) to macro scale (systems) will be illustrated with an example in the third chapter.

MATERIALS

The properties of fine grained soils such as silt or clay considerably influence the migration of water. Especially their small pore sizes, their platy habit, and their high specific surface area generally lead to very low hydraulic conductivities. Therefore, it is indispensable to accurately determine these properties for a reliable assessment of the hydraulic conductivity. Fine-grained soils are soils with a grain size distribution ranging from 0,0002 to 0,2 mm, i.e. soil textures from clay and silt up to fine sand. The hydraulic conductivity, K_f, of these materials is generally smaller than 10^{-4} ms^{-1}. In the following the most important physical, chemical and dynamical properties of fine grained materials that affect the hydraulic conductivity will be explained in detail and their measurement methodologies will be illustrated.

Material properties

Generally the hydraulic conductivity depends on the soil matrix, the type of the soil fluid (density and viscosity), and the relative amount of soil fluid (saturation) present in the soil matrix. In this chapter we will focus on the important properties relevant to the solid matrix of fine grained soils which include the texture and fabric and its mineral phase content.

Physical Material Properties

Soil Density

In soil science one can distinguish between the soil particle density ρ_p which is the mass of the soil particle m_p per volume of the particle V_p and the so called bulk density which is the ratio of the mass to the bulk volume V of a given amount of soil. For the latter two different definitions depending on the scientific discipline or the application are possible, i.e. the wet bulk density ρ_{wb} which is the soil mass m_s, plus the mass of the including water m_w, per unit volume V and its dry bulk density ρ_{db} which is the mass of oven-dry soil per unit volume of moist soil.

$$\rho_p = \frac{m_p}{V_p}$$
(1)

$$\rho_{wb} = \frac{m_s + m_w}{V}$$
(2)

$$\rho_{db} = \frac{m_s}{V}$$
(3)

Soil Texture (Grain Size and Grain Size Distribution)

Mineral soils are mostly classified according to grain size and grain-size distribution (grainsize fractions and grading) also generally summarized as the soil texture. A major factor that influences the rate of water flow through soils is the size of the particles. It can generally be stated that the smaller the particles, the smaller the voids and the stronger the flow resistance. This explains the very low hydraulic conductivity of fine-grained materials and their preferred use as sealing elements for example in buffers and backfills of underground storage facilities or landfills. The grain-size distribution gives a good picture of the content of clay minerals since they appear almost entirely in the clay fraction (< 2 μm). The compaction properties of the material can be estimated by the character of the entire grain size curve, from which one can conclude, if the material has very low hydraulic conductivities and can be used for water sealing purposes (Pusch, 2002). Whereas the grain-size distribution of the coarse fractions (gravel and sand) can be determined by sieving, the fine fractions (silt and clay) must be determined by sedimentation of the dispersed soil.

Soil Structure (Porosity and Pore Size Distribution, Geometry and Shape of the Pores, Tortuosity)

The term structure of soils generally refers to the pore geometry. The pore volume or porosity is a measure of the void spaces in the soil and is a fraction of the volume of voids over the total volume, between 0-1 or as a percentage between 0-100 %.

$$\Phi = \frac{V_V}{V}$$

(4)

where V_v is the volume of void-space and V is the total or bulk volume.

Porosity of surface soil typically decreases as particle size increases. This is due to soil aggregate formation in finer textured surface soils when subject to soil biological processes. Aggregation involves particulate adhesion and higher resistance to compaction. The transport of water, solutes, and gases in soil is not only influenced by the absolute size of the pore volume but also on the nature of how the pores are connected which is summarized under the term tortuosity. In soil the tortuosity is closely related to soil surface area and the pore-size distribution. Both porosity and tortuosity of fine-grained soils are considerably small due to the plat-like shape of the particles.

Chemical Material Properties

Composition of the Mineral Phases

Minerals in natural soils originate from degraded rock, most of them belonging to the silicates, sulphates, sulphides, and carbonates. Especially the clay minerals show a platy habit leading to a very high surface area to mass ratio and have a considerable influence on the hydraulic conductivity. The distinction of minerals is mostly based on crystal structure and chemistry. Their crystal structure is responsible for a number of their characteristic chemical properties such as cation exchange capacity or high sorption capacity (Pusch, 2002).

Hydraulic conductivity of swellable clayey and clay-enriched silty soils strongly depends on the density, the type of the adsorbed ions and the salinity of the percolating water (Scheffer, 1992). For example it is the high swelling properties that provide sodium bentonite's unique sealing qualities. As the clay hydrates and swells, the path for water to flow through becomes complex as the clay platelets intersperse. The large fraction of interlamellar, immobile water in smectites yields a much lower hydraulic conductivity than of soils with other minerals at any bulk conductivity. Thus, clays with micas, illites, and kaolinites as major minerals are about 100 to 100000 times more conductive than smectite in montmorillonite form at one and the same void ratio (Pusch, 2002). (See figure1)

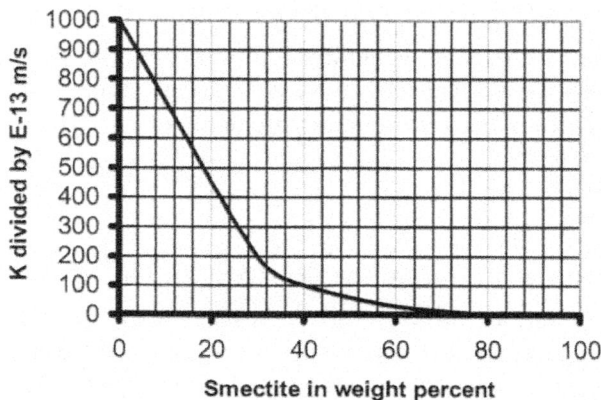

Figure 1. Approximate relationship between smectite content and hydraulic conductivity (Pusch, 2002).

The type of absorbed cations of the clay is also important with respect to the hydraulic conductivity since bi- and polyvalent cations cause growth in the stack thickness and size, which means that the voids between the stacks of lamellae and thus the hydraulic conductivity are bigger in the Ca than in the Na montmorillonite (Pusch, 2001). Also the electrolyte concentration has a substantial influence on the interparticle distance at low and moderate densities and thus on the hydraulic conductivity because the stacks of lamellae that form a network with rather much space will coagulate at high electrolyte concentrations.

Specific Soil Surface

The magnitude of the specific surface area of a soil depends largely on the amount of clay and type of clay minerals in the soil. The specific surface area differs largely between types of clay minerals (Table 1). Especially in fine-grained materials one can distinguish between an external and an internal surface, the latter being the interlamellar space of the minerals. The total specific surface area is a factor that can relate grain-scale properties to macro-scale physical and chemical properties of a porous medium. Large specific surface areas lead to much interaction of ions and water molecules with the soil particles. Therefore the total surface determines many physical and chemical properties of the soils (Petersen et al. 1996). In porous media filtration theory, a nonlinear relationship between saturated hydraulic conductivity and surface area has been established for structureless, randomly packed, noncompressible particles (Kozeny, 1927; Carman, 1937; Grace, 1953), the so-called KozenyCarman relationship.

Table 1. Typical specific soil surface area data of clay minerals (Pusch, 2001).

Mineral	Specific surface area, m²/g	
	External	Internal
Kaolinite	10–15	10–15
Illite	100–150	100–150
Smectites	100–150	800–1000

Dynamical Material Properties

Matric Potential

Matric potential refers to the tenacity with which water is held by the soil matrix and, in the absence of high concentrations of solutes, is the major factor that determines the availability of water to plants. Differences in the value of matric potential between different parts of the soil also provide the driving force for the unsaturated flow of soil water after any differences in elevation have been allowed for (Mullins, 1991).

The total potential ψ_t of soil water refers to the potential energy of water in the soil with respect to a defined reference state and can be divided into three components:

$$\psi_t = \psi_p + \psi_g + \psi_o$$

$$(5)$$

with ψ_p as pressure potential, ψ_g as gravitational potential and ψ_o as osmotic potential. The sum of gravitational and pressure potential is called the hydraulic potential ψ_h. Differences between the hydraulic potential at different places in the soil provide the driving force for the movement of soil water. Matric potential ψ_m is a subcomponent of pressure potential and is defined as the value of ψp where there is no difference between the pressure of air or gas in the soil and the gas pressure on the water in the reference state (Mullins, 1991).

The relationship between water content and soil water potential (capillary tension) is described in the water retention curve. The curve is characteristic for different types of soils and is also called soil moisture characteristic. It also depends on the geometry and network of the pores. Fine-grained soils show high residual water contents and high changes in capillary tension are necessary that they release the water.

The retention curve shows a hysteresis which means that depending on the history of the soil with regard to watering and drainage, the shape of the curve is different, a behaviour which is explained by the 'ink-bottle model' (Schuhmann, 2002).

For the analytical description of the water retention curve there are different approaches. Brooks and Corey (1964) combine the mathematical θ/ψ–relationship with the conductivity model of Burdine (1953), but the most common approach is the one from van Genuchten (1980) combined with the conductivity model of Mualem (1976a) which allows a direct determination of the hydraulic conductivity by numerical simulation. Up to now it was state of the art to determine the capillary tension by tensiometers and to determine the moisture with the θ/ψ–curve, yet the influence of the hysteresis does not allow distinct results. Therefore we propose other soil moisture measurement methods to derive soil hydraulic conductivity.

Moisture Content

As with density there exist several different definitions of water content or moisture. The gravimetric water content can, similarly to the bulk density, be expressed on a dry mass, θ_{db}, or wet mass basis, θ_{wb}, and gives the ratio between the mass of the porewater and the mass of the dry solid substances resp. the solid substance plus mass of water. It can be expressed in percent units

$$\theta_{db} = \frac{m_w}{m_s} * 100$$

(6)

$$\theta_{wb} = \frac{m_w}{m_s + m_w} * 100$$

(7)

The volumetric water content can also be expressed on a volume basis as the ratio between the volume of water and the total volume of the soil sample:

$$\theta_{vb} = \frac{V_w}{V_t} * 100$$

(8)

Since the soil water characteristic from the θ/ψ–curve especially for fine-grained soils is ambiguous, its transformation into a ψ/k_f-curve for the determination of the hydraulic conductivity is ambiguous too. Therefore, instead of the capillary tension, in the following the moisture content will be used as the relevant parameter for a more precise determination of the hydraulic conductivity. For fine-grained soil samples, the wide distribution of void size means that the various pore water components play different roles. Determination of their relative amounts requires heating to different temperatures. The water in the large voids is lost at about 100°C, the water in the fine capillaries at 105°C or slightly more, while the hydration shell of interlayer cations in swellable clay minerals is lost at temperatures appreciably higher than 105°C. Although the determination of the water content by thermal analysis is a very accurate method and mostly used as a reference, this method is destructive and non-operational. But fortunately the range of possible soil moisture determination methods has increased considerably since the beginning of the eighties of the last century. Especially the electromagnetic moisture measurement methods are promising new techniques for a reliable and in-situ determination of the soil moisture and thus of the soil moisture/ hydraulic conductivity-relationship in the field. One of these new techniques will be presented in the following chapter.

MEASURING METHODS

Physical Methods

Bulk Density

The methods available for the measurement of soil bulk density fall into two groups. In the first are the long established direct methods, which involve measurement of the sample mass and volume (core sampling, rubber ballon, sand replacement, clod). In the second group the attenuation or scattering of nuclear radiation by soil is used in conjunction with a calibration relationship to give an indirect measurement of bulk density (Mullins, 1991).

Soil Texture (Particle Size Distribution, Grading Curve)

The particle size distribution analysis is one of the most principle determinations in soil science and its knowledge already allows relatively good estimations of soil hydraulic properties. Grain size distribution is very important for the bulk density.

Principally with particle size distribution analysis there exist two problems: (1) from the wide range of particle sizes it follows that the analysis cannot be carried out by one single method alone and (2) since the particles show different stabilities it is nearly impossible to exactly distinguish between primary particles and aggregates, the latter being valid especially for fine grained materials such as clay and silt. Therefore a chemical and/or physical pre-treatment of the soil sample is indispensable to minimize aggregation of particles. The separation of the different particle sizes is carried out be sieving with exactly defined mesh sizes. The fine grained fraction, i.e. particle ranging from <63 μm to 2 μm, are normally separated by so called sedimentation analysis, using the different sinking properties of the grains in liquids (Stokes' law). This law is applied in several methods like aerometer according to Casagrande, Andreasen Pipete, Koehn Pipete, Atterberg or Kopecky, that yield the so called grading curve. This method reaches its limitations when the clay breaks up into fine, medium, and coarse clay. For determination of the clay fractions sedimentation is speed up by centrifugation applying several times of earth gravitational force.

All these methods are time consuming and determine size distribution of spherical equivalent particles as Stokes' law is based on the assumption of spherical particles. Faster methods of particle size distribution apply laser light scattering, X-ray absorption, acoustic spectrometry or dynamic light scattering. A systematic comparison of advantages and disadvantages in relation to particle size and particle shape was given by Latief (2010). The shape of platy clay minerals has a strong influence on their sedimentation behaviour and thus influences the determined equivalent particle size. Therefore, some methods allow introduction of a shape factor (e.g. Konert & Vanderberge, 1997) for calculation of particle size distribution.

Soil Structure (Pore Distribution)

Beside the calculation of the porosity Φ from the density values, it can also be measured directly by an air pycnometer which is based on the Boyle-Mariotte law ($p_1 V_1 = p_2 V_2$).

The relationship between the decrease of pressure (from p_1 to p_2) and the pore volume in the sample must be taken from calibration curves. This yields the air-filled pore volume V_A. To extract the total volume V, an additional water content determination to determine the volume of water V_W in the sample is necessary.

$$\Phi = \frac{(V_A + V_W)}{V}$$

(9)

Specific Surface Area

The determination of the specific surface area helps to identify expandable minerals and to estimate the degree of cementation of expandable clays. The measurement is carried out by determining the external and internal surface areas. The external surface area is measured by the adsorption of non-polar gases like nitrogen which cannot enter the interlamellar space. The internal surface is determined by measuring the total surface using water and subtracting from it the separately determined external surface (Pusch, 2002).

Water absorption capacity according to Enslin/Neff describes the property of soil at 60°C dried up to weight constancy to absorb water by capillary action and to retain it. It mainly depends on content and type of the clay minerals. The Enslin-Neff values are known to be determined by the amount of exchangeable Na⁺ and to a lesser extent by the smectite content (Neff, 1959). Hence the Enslin-Neff method provides an indirect estimate of the Na^+ content. Evaporation during the test has to be restrained as it is known to have a serious affect on the water-uptake capacity values measured by the Enslin-Neff device (Kaufhold & Dohrmann, 2008). Nevertheless, the test provides an index value which gives some indication of the size of the specific surface and activity of the fine grain.

Chemical and Mineralogical Methods

Phase Analysis

Identification and quantification of mineral phases is performed by X-ray diffraction analysis and analysis of X-ray diffraction patterns from powdered samples by Rietveld method or pattern summation methods supported by complementary analyses (e.g. CEC, XRF, STA) (Omotoso et al., 2006, Emmerich, 2011).

- *X-Ray Diffraction Analysis*: X-ray diffractometry (XRD) is the standard analysis in mineralogy providing rapid information on clays and non-clay minerals present in a powder sample. An X-ray beam of defined wavelength (e.g. Cu Kα) is diffracted at the lattice of each mineral. According to Bragg's law the diffracted beam is commonly recorded at angles between 2-80° 2θ. The intensity ratio of peaks for each mineral depends on many factors (like chemical composition, preferred orientation and others, see e.g. Moore & Reynolds, 1997). Identification of the phases is made according to the JCPDS register (International Centre for Diffraction Data; JCPDS 1983) incorporated in commercial software. Identification of clay minerals often requires additional XRD analysis of so-called textured samples. The sample is slurried by appropriate chemical and/or physical treatment and the >2 μm fraction separated in a centrifuge and discarded. The remaining suspension (< 2 μm) is placed as drops on a slide or sucked on top of a ceramic disc whereby the clay mineral orientates during the sedimentation and drying process more or less on a parallel basis (texture) (Moore & Reynolds, 1997). Subsequent solvation with ethylene glycol will identify the swellable clay mineral phases. Calcinating at 550°C destroys the existing kaolinite through dehydroxilation allowing chlorite to be identified (http://pubs.usgs.gov/of/2001/of01- 041/htmldocs/flow/index.htm).

- **Cation Exchange Capacity (Methylene Blue and Cu-Trien methods)**: Determination of cation exchange capacity (CEC) is performed by homoionic saturation of exchangeable positions by an Index cation (Dohrmann & Kaufhold, 2010) and indicates the amount of smectites in fine grained materials. Common index cations are methylene blue (MB), Ba^{2+}, NH_4^+ or copper triethylenetetramine(Cu-Trien). Despite MB is widely used (VDG P 69 1988) the method suffers certain restrictions that are discussed in detail by Kahr and Madsen (1995). If the CEC is determined by MB of a sodium exchanged material at neutral pH it would result in similar values as determined with the ammonium (NH_4^+) acetate method. MB can even be used for determination of SA if montmorillonite surface area per charge corresponds with the area of the MB of 130 $Å^2$, i.e. the interlayer charge of the montmorillonites must amount to 0.28-0.33 charges per half unit cell.

- Occupation of exchangeable positions in natural state of clay containing is of particular significance for water adsorption, swelling and resulting microstructure and thus permeability of clays especially bentonites. Exchanged cations can be determined by AAS or ICP-OES from supernatant of CEC determination if precaution is taken to prevent dissolution of soluble minerals (Dohrmann & Kaufhold, 2010). According to MüllerVonmoos & Kahr (1982) the cations bound to the exterior and interior surfaces of montmorillonite can be exchanged for an alcoholic ammonium rhodanide solution in the presence of soluble salts, too. The ion distribution of exchangeable bound ions (Ca^{2+}, Mg^{2+}, Na^+, K^+) is of particular significance for water absorption and swell capacity, the microstructure and thus the water permeability of bentonite.

- **Swell Capacity Method**: The ASTM D 5890 method id used to determine the swell capacity of bentonite. Two grams of dried (105°C) pulverised sample is placed in 90 ml of de-ionised water in standard 100 ml glass cylinders. Of the sample, 0.1 g at a time is sprinkled into the water within 30 seconds until the whole amount has been used up. The cylinder is filled to 100 ml with de-ionised water and temperature is measured. After 24 h, the swell volume is determined in ml ignoring any flocculated material.

- **Fluid Loss Test**: The fluid loss test is a very reliable test regarding the permeability behaviour of bentonites. It enables the evaluation of fluid loss properties of a clay mineral film deposited on a filter paper from a 6% solids slurry of clay mineral at 100 psi (690 kPa) pressure as a measure of its usefulness for the permeability of hydraulic conductivity reduction. This method was adapted from American Petroleum Institute drilling fluid specifications for bentonite.

Dynamical Methods

Matric Potential

Namely the tensiometer and the pressure-membrane (or pressure-plate) apparatus, that either measure, or generate, the matric potential as defined above are used. These instruments measure the difference in pressure across a semipermeable 'membrane' in contact with the soil on one side and the soil solution (i.e. the equilibrium dialysate) on the other. The 'membrane' is permeable to solution but not to solids (Passioura 1980).

Principally the tensiometers are suitable for the determination of the hydraulic conductivity in the field and a measurement precision of ± 1.0 vol.-% is principally sufficient, yet there can occur big uncertainties when transforming the capillary tension into moisture or hydraulic conductivity due to described hysteresis effect. Also maintenance and calibration of the tensiometers especially in fine-grained soils are quite complex since they tend to run dry very quickly due to the high water suction of finegrained soils. Also the measurements are rather punctual. However they can be used as a reference method for our purpose.

Moisture Content

Since moisture content is a decisive criterion of many porous materials, there exists a broad variety of different methods. Generally one can distinguish between direct methods, where the moisture is determined directly by physical or chemical methods, or indirectly by determining a property that is mainly a function of the water content.

The most common direct method is the thermogravimetric method, where a sample of the soil is dried at 105°C to a constant weight. Other direct methods use e.g. calcium carbide, sulphuric acid, or phosphorus pentoxide, which react with the water present in the sample. Because all these methods require sampling and considerable laboratory equipment, their use for soil moisture measurement in the field is not very practicable. Here the indirect methods are more common and more differentiated using mostly physical parameters such as electrical, radiometric, acoustic or thermal soil properties. A good survey is given in Schmugge et al. (1980).

In this chapter we will focus especially on the electromagnetic methods that use the dielectric properties of the soil. The principle is based on a functional relationship between the dielectric permittivity of the soil and its volumetric water content. Different methods take advantage of the high relative permittivity of water ($\varepsilon_r = 80$) compared to that of dry soil ($\varepsilon_r = 3\text{-}5$). One of the most well-known meanwhile is the Time Domain Reflectometry (TDR) (Topp et al. 1980), where the transit time t of an electromagnetic pulse on a wave guide of length l, which is buried in the soil, is measured. The relative permittivity ε_r is determined according to:

$$\varepsilon_r = \left(\frac{c_0 t}{2l}\right)^2$$

(10)

with c_0 as the velocity of light in free space. Relating the measured ε_r of different soil samples to the volumetric water content determined by the thermogravimetric method, allows to establish a so called calibration function. One of the most well-known calibration functions is the Topp-polynomial (Topp et al. 1980) which yields the volumetric water content θ_v according to:

$$\theta_v = -5.3 * 10^{-2} + 2.92 * 10^{-2}\varepsilon_r - 5.5 * 10^{-4}\varepsilon_r^2 + 4.3 * 10^{-6}\varepsilon_r^3 \quad (11)$$

Conventional TDR-sensors are normally fork-like metallic wave guides of several tens of centimetres that penetrate the soil, giving a rather punctual measurement. Yet the determination of hydraulic conductivity on a field-scale basis based on soil moisture measurement requires more large-scale sensors why a flat-band-like TDR-cable sensor called TAUPE has been developed (Brandelik & Huebner, 1999). Due to a plastic coating of the copper wave guides this sensor is capable of sensing up to 30 m of the surrounding soil. Both an integral soil moisture value and a moisture profiling along the length of the buried cable according to a new TDR inversion technique can be accomplished (Schlaeger, 2005).

PROCESSES

Dynamic of the Surface Moistening

The moistening of surfaces obeys certain natural laws which were established during the last 200 years. The development was started in the beginning of the 19th century by several scientists from the fields of physics and chemistry. In this chapter we will exemplify some important laws with respect to surface moistening.

Laws

Young-laplace Equation

In 1805 Thomas Young and Pierre-Simon Laplace both described independently from one another a fundamental equation with respect to interface science. The Young-Laplace equation describes the correlation between surface tension, pressure and surface curvature of a system consisting of two phases. Such a system could be e.g. a liquid drop on a solid surface or a liquid in another immiscible liquid. Surface tension of a liquid results from attractive interaction of the liquid molecules. A molecule located within a liquid is surrounded by other molecules, so the resultant force is zero. This does not apply to a molecule at the surface, since a part of interaction is missing at this place. The molecule is bordered by air molecules on the upper side and these intermolecular forces are of weak nature. This leads to an inward looking force. The energy required to overcome this force is the surface tension, sometimes also called surface energy.

Lucas-Washburn & Modified Lucas-Washburn

The predefined aim of the studies of Lucas (1918) and Washburn (1921) was to develop a theoretically established law, which determined the capillary head existing in an arbitrary capillary system, as a function of time. The first approach was to immerse a wettable cylindrical tube vertically into a solution. The surface tension of the liquid becomes noticeable as the length of the cross section ($2*r*\pi$) multiplied by the surface tension (σ), perpendicular to the direction of the tube. This force elevates the liquid to a height where it is equilibrated by the gravity.

$$2r\pi\sigma = r^2\pi m_s h_0$$

(12)

where h_0 is the maximum pressure head and m_s the specific mass. Thus the maximum height entirely depends on the surface tension, on the radius of the tube and on the specific mass of the solution measured. The penetration speed of the liquid due to the pulling force diminishes with the height because the mass of the liquid increases. Moreover the rise of the liquid is slower the tougher the liquid is. After the viscosity of the solution has been taken into account (Poiseuille) and assuming that wetting isn't complete, the Lucas-Washburn equation for the capillary rise is

$$h^2 = \frac{r\sigma \cos\theta t}{2\eta}$$

(13)

Here θ is the contact angle, t is the time and η is the viscosity. The contact angle is between $0°$ and $90°$, so cos(θ) lies between 0 and 1.

Methods

Contact Angle Measurement

The basis of the contact angle measurement goes back to Thomas Young (1805), who related the contact angle to the surface tension:

$$cos\ \theta = \frac{\sigma_s - \sigma_{ls}}{\sigma_l}$$

(14)

θ is the contact angle, σ_s the surface free energy, σ_{ls} the solid-liquid surface energy and σ_l is the surface tension of the liquid. We can distinguish between three cases relating to the contact angle. If $\theta < 90°$, the liquid wets the solid surface, if $\theta > 90°$, the sample is hydrophobic and the liquid doesn't wet it or wets it only partially and if $\theta = 0°$, the solid surface is totally wettable, the liquid spreads over the surface.

In practice the liquid drop is put on the solid surface, which has to be as straight as possible, plane and also clean. A light source, which is positioned in the rear lets the drop appear dark. θ can be measured directly using a goniometer or with the help of an optical calculating system which employs the equation of Young-Laplace. The goniometer measuring leads to a relatively large error ($\pm$ 2%) and is not applicable for small angles and irregular contact lines (Dimitrov et al., 1991). In case of small drops the hydrostatic effects can be neglected and the contact angle can be calculated from the height of the drop (Butt et al., 2006).

Dynamic Contact Angle Measurement

The processes happening at the solid-liquid interface during wetting and dewetting are best described by the dynamic contact angle. The interface at the contact between liquid drop and solid surface doesn't appear suddenly, but it needs a certain time until a dynamic equilibrium is reached. In practice the measuring of the dynamic contact angle works in the way that a liquid drop is spread on the solid surface and then extended by means of a needle. The solid-liquid interface migrates outwards and the contact angle can be measured by defining certain degrees steps. Studebaker & Snow (1955) developed an equation for the determination of the dynamic contact angle.

$$cos\ \theta_2 = \frac{\sigma_1 \eta_2 t_1}{\sigma_2 \eta_1 t_2}$$

(15)

The authors determined dynamic contact angles of powder samples by measuring the time required for a liquid to imbibe the powder bed. This time was then compared to a reference sample with cos θ = 1 (contact angle = 0°). This method assumes that the differences in the penetration rate are due only to differences in contact angle, after taking surface tension and viscosity into account (Yang & Zografi, 1986).

Capillary Rise Method

Jones & Ray (1937) investigated the determination of the surface tension of water and several salt solutions. They developed a differential method to determine this property of liquids. The experimental set-up of the capillary rise method consists of a tight cylindrical tube and a broad tube being connected with each other. The vertical level difference between the meniscus in the tight tube and the extended one has to be measured. The density of the liquid, which also needed, may be determined directly by the use of a hydrometer.

$$\sigma = \frac{rhg(D-\beta)}{2} \cos \theta$$

(16)

r is the radius of the tight tube measured at the height of the meniscus, h the capillary rise, g the acceleration of free fall, D the true density of the liquid, β the density of the gas phase (air plus water vapor) at the temperature and the barometric pressure when the experiment is made and θ is the contact angle. θ should be zero in glass and silica tubes if the tubes are clean (Jones & Ray, 1937). First the elevation between the lowest levels of the menisci must be read off to get the approximate value of the capillary rise. This value has to be corrected for the liquid by means of the Rayleigh formula. Jones & Frizzell (1940) for their part examined the influence of the concentration of the solution on the capillary rise. Therefore they used diluted salt solutions of different concentrations. The most noticeable feature of the results was that the penetration height of water was higher than those of the diluted solutions. This was interpreted as an evidence for a higher surface tension of water compared to the salt solutions. Measurements based on Washburn's equation do not only depend on the particle size but also on the pore size distribution. Addition of fine particles to the measured bed increases the penetration rate of liquid and improves precision of the measurement (Dang-Vu & Hupka, 2005).

Wilhelmy-Plate

The Wilhelmy-Plate method is utilized to determine the surface tension of a liquid. It can also be used in order to study the contact angle during capillary rise. In doing so a plate is contacted with the surface of the examined liquid where a meniscus forms at the contact point of the two phases. Due to this meniscus a force between the phases appears which originates from the wetting. By pulling the plate upwards a force (surface tension) manifests itself.

Sessile Drop

The interface science makes use of different methods with regard to measure both, properties of liquids (e.g. surface tension) and properties of solids (e.g. static/dynamic contact angle or surface energy). The sessile drop method is an example for a measurement on a solid. For this purpose a drop of liquid (in most of the cases a reference solution is used) is spread on a solid surface and the static contact angle of the liquid is measured optically. Bachmann et al. (2000) developed a sessile drop method by modifying Young's equation on two points, since it is strictly applicable only to completely uniform and plain surfaces:

1. A correction factor was introduced, which is defined as the ratio between the actual and the apparent area. This leads to the equation of Wenzel

$$cos(\theta_{obs}) = r \, cos(\theta) \tag{17}$$

Eq. (17) was developed due to the fact that the observed contact angle is smaller than the ideal (intrinsic) angle as long as this is below 90° and larger if the intrinsic angle is above 90°. So, the precision of the contact angle measurement therefore depends on the magnitude of it.

2. The Cassie-Equation is considered as an empirical approach describing the apparent contact angle on a chemically heterogeneous surface

$$cos(\theta_{obs}) = f_1 \, cos(\theta_1) + f_2 \, cos(\theta_2) \tag{18}$$

Fluid Flow

Hydrology

According to the physical law of conservation of mass the equation of water balance is valid. The input precipitation (N) is equal to the sum of the current total evaporation (Haude, 1958) ETa (evaporation of soil, transpiration of plants and evaporation of interception), the sum of runoff (surface runoff Q_O, lateral runoff Q_L and leaching Q_V) and change of water content in soil layers within the observation period ($\Delta_\theta = W_A - W_E$).

$$N = ET_a + (Q_O + Q_L + Q_V) + (W_A - W_E) \tag{19}$$

The description of water balance in vertical soil profiles is the fundamental requirement for the description of water balance in areas. Numerous investigations on the regionalization of point methods of measurement nowadays provide well-founded transmission options. The following sections consider the water balance of a vertical profile.

Analytical Basics

The major focus of the investigation lies on the flow of a solution (in this case water) through the soil matrix. The water flow not only underlies gravity, but is also influenced by the soil properties (Hillel, 1980). Driving forces like gravity (hydrostatic forces), adsorption, cohesion, osmotic forces due to dissolved salts etc. cause water movement through their resultant (Czurda, 1994). Concerning the water movement we distinguish between advective flow and diffusion. Diffusion is irrelevant in the case of materials owing high hydraulic conductivity. The mathematical expression of the potential can be used to describe the flow of water through soil. It should be noted that the theory of the potentials considers neither the geometry of the pore space nor the mechanisms of water binding. These are included in the matric potential. The mechanical energy is taken into account but not the thermal energy. The stationary flow within the unsaturated zone is described by Darcy/Buckingham (equation of motion). Darcy's law features flow by means of a unit volume in dependency of the hydraulic conductivity of soil and in dependency of a potential gradient. The flow velocity depends on the hydraulic conductivity (kf-value) of the soil and on the total potential (Ψ). k_f depends on the water content θ.

$$v_x = -k_{fx}(\theta) \cdot \frac{\partial \Psi}{\partial x} \qquad v_y = -k_{fy}(\theta) \cdot \frac{\partial \Psi}{\partial y} \qquad v_z = -k_{fz}(\theta) \cdot \frac{\partial \Psi}{\partial z} \tag{20}$$

k_f does not vary linearly depending on water content, but it follows a relationship which is characteristic to each soil. The hydraulic conductivity decreases with the square of the capillary radius. The air in the soil is considered to be stationary. In case of transient conditions Darcy's law (equation of motion) is combined with the equation of continuity (validity of conservation of mass).

$$\frac{\partial v_x}{\partial x} + \frac{\partial v_y}{\partial y} + \frac{\partial v_z}{\partial z} = -\frac{\partial \theta}{\partial t} - S \tag{21}$$

S represents a term containing a sink or a source. Combining (20) with (21) considers the change in water content during water flow. This equation is known as the partial differential equation of unsaturated flow (unit s^{-1}) within the soil matrix.

$$\frac{\partial \Psi}{\partial t} \cdot C + S = \frac{\partial}{\partial x}\left[k_{fx}(\theta) \cdot \frac{\partial \Psi}{\partial x} \right] + \frac{\partial}{\partial y}\left[k_{fy}(\theta) \cdot \frac{\partial \Psi}{\partial y} \right] + \frac{\partial}{\partial z}\left[k_{fz}(\theta) \cdot \frac{\partial \Psi}{\partial z} \right] \tag{22}$$

C ($=d\theta/d\psi$) is defined as specific moisture capacity of the soil. Richard's equation contains the relationship between water content (θ) and soil water tension (Ψ, set equal to the potential) and also the relationship between hydraulic conductivity and water content. These relationships are extremely non-linear. Therefore, the solution of equation (24) (calculation of water and solute transport) requires numerical methods (Jentsch, 1992; Philip et al., 1974). It is foreseeable that we need to know two parameters in order to describe the water movement within the soil: total potential (Ψ, represented by the soil water tension under described boundary conditions) and water content.

Soil Mechanics and Soil Hydraulics

Soils store and transport water within their pore system. This property is determined by the hydraulic conductivity and the texture of the pores. The hydraulic conductivity depends on the water content of the soil. Considering an unsaturated "ideal" soil with a uniform microstructure without macropores the following relationships are relevant:

- Ratio volumetric water content to soil water tension also called the soil moisture characteristic or pF-curve
- Ratio soil water retention to hydraulic conductivity
- Ratio volumetric water content to hydraulic conductivity

The soil water tension equals to the sum of bonding forces performed by the soil matrix with regard to the water. The pF-curve shows a characteristic course for each different soil (Scheffer, 1992). It depends on the particle size distribution, the formation and crosslinking of the pores and also of their size. The proportion of organic matter and the chemical composition of the wetting phase also affect the pF-curve. The water holding capacity of sand is low (if pF>4.2) the residual saturation θ_r equals a water content less than 3 vol. %), so sand releases water situated within the pore volume at low soil water tension differences (low specific water capacity C, that means a slight gradient of the tangent with respect to the pF-water content-curve at $0 < pF < 2.5$). In contrast, clay shows a high residual saturation (if pF = 4.2, water content equals ~ 30 vol. %) and the release of water needs high changes of the soil water tension (high specific water capacity C, that means high gradient).

Depending on the history of the soil with respect to watering and dewatering the shapes of the pF-curve for a single soil are different, an effect called hysteresis. It is assumed that the capillaries which connect the pores show a smaller cross section than the pores themselves. If the soil water tension is high, the convex meniscus holds on to the upper border of the pore (by means of retention forces). This soil water tension is higher than necessary in order to move the concave meniscus (formed by wetting resistance) into the capillary in case of watering. The main influencing factors of the pF-curve are grain size, soil structure and hysteresis.

There are manifold approaches for the analytical description of the pF-curve. The approach of Brooks & Corey (1964) combines the mathematical ratio of θ to Ψ with the conductivity model of Burdine (1953). The definition of effective water content, also called relative saturation index or soil-water-retention, is

$$Se = \frac{\theta - \theta_r}{\theta_s - \theta_r}$$
(23)

with θ as current water content, θ_r as residual water content (simplified $= 0$, easier to measure at pF $= 4{,}2$) and θ_s as water content at saturation (measured at pF $= 0$), conform to porosity. Besides the approach of Campbell (1974) the common approach originates from Van Genuchten (1980) and Mualem (1976a), VGM in the following. The VGM approach prevailed in the literature and will be considered subsequently. Therefore we combine the (Ψ)-relationship with the conductivity model from Mualem. Unlike the model of Brooks & Corey-Burdine (Berger, 1998) this approach doesn't take any sharp air inlet into account. Thus this approach is solvable from the analytical point of view if we consider certain boundary conditions

$$Se = \frac{\theta - \theta_r}{\theta_s - \theta_r} = \left[\frac{1}{1 + (\alpha h)^n} \right]^m$$
(24)

α, m, n parameters of shape, dependent on grain distribution of the soil:

$\alpha = 1/h_b$, ; h_b = soil water tension (ψ) at air inlet point,

$m = \lambda/(\lambda + 1)$; λ = index of pore size,

$n = \lambda + 1$; hence $m = 1 - 1/n$

h soil water tension (ψ) at the water content θ

So the description of the hydraulic conductivity/soil moisture-relationship becomes possible from the analytical point of view based on the prediction model of Mualem:

$$\frac{k_f(\theta)}{k_s} = Se^{\gamma} \cdot \left[1 - \left(1 - Se^{\frac{1}{m}} \right)^m \right]^2$$
(25)

$k_f(\theta)$ hydraulic conductivity at the current water content

k_s hydraulic conductivity at water saturation. k_s is often set constant, but also depends on the texture of the soil

γ takes the influence of tortuosity into account, usually $\gamma = 0.5$.

By means of inverse identification of parameters (Schultze et al., 1996) the approximation of the values from numerical simulations to those from field trials is possible. Thus the sought parameters are provided directly. The measuring of water tension by means of tensiometer represented until now the state-of-the-art. The water content of the soil was determined via pF-curve (with knowledge of the water tension) and furthermore the k_f-value. Until now no exact results could be achieved due to the influence of hysteresis.

Soil Moisture

In order to solve the Richards equation (22), knowledge of properties of the soil is required. Until now these properties were taken from the fundamental pF-curve. There the pF-value represents the common logarithm taken from Buckingham's "capillary potential" expressed as cm of the water column. The shape and uniqueness of the pF-curve depends on the texture. The texture includes pore size distribution (structure) as well as storage and arrangement of the particles (texture). For analytical modeling further parameters are necessary. These parameters describe the water content of the soil at different boundary conditions. Alongside the natural saturation water content (θ_s) where pF $= 0$, the absolute saturation (θ_s^*) is also an important parameter. θ_s^* correlates to the porosity of the soil. Absolute saturation cannot be achieved by rewatering because -according to structure and texture- certain parts of the pores remain air-filled. According to extensive investigations the following equation is valid

$$\theta_s = 0.8 \div 0.95 \, \theta_s^*$$

(26)

Thus soil is liable to be seen as three-phase system. A further parameter is the remaining water or residual saturation defined as θ_r, where the aqueous phase is not constant anymore. The associated pF-value is 4.2. The distribution of the three components water, air and soil is described by various "mixing models". The aqueous phase is distinguished between "free water" and "bound water" that is adsorbed on or within the particles. The electrostatic forces surrounding the solid (here: soil particle) act outwards. These forces result from molecules that are not compensated electrical all-round. The wetting property of a solid with regard to water depends on the strength of these forces. If the cohesion forces of the water molecules are less than the surface forces, the water molecules absorb on the surface. Bound water is able to absorb further water molecules by means of associate forces, however, this binding is not stable (Huebner, 1999).

Bound water prefers ionic bonds. The surfaces of fine-grained materials such as clays are saturated by ions. The sorption forces between ions and the surface of the clays are greater than the non-polar sorption forces between surface and water molecule. As a result the wetting property of clays increases and a hydrate envelope around the metal cations is established. Also crystal water, i.e. water bound within the lattice of the soil particles, is present and water can condensate within the capillaries. If two water films get into touch, the water molecules flow together and form carrying menisci within the soil pores and more water molecules are attracted by the surface tension. If the soil air is saturated with vapor, the water condenses above the concave meniscus. At this place the vapor pressure is smaller than above the convex or the flat meniscus. The molecular forces get saturated by steam or by liquid water.

The main part of the water in the soil is not influenced by molecular forces. It has zero potential and underlies gravity. The water contents θ, θ_s and θ_r have been determined in on lab-scale depending on soil water tension. Because of the hysteresis it was distinguished between the watering and dewatering of the sample. The reasons for the hysteresis of the pF-curve are that the advancing contact angle between soil matrix and soil water is greater than the retreating contact angle, the effects regarding the geometry of pores, water bound on clay mineral surfaces, and enclosed air. Statements leading to a reliable approximation of the hydraulic conductivity are therefore impossible. The extension of uncertainty by 70 times results from empiric measures (Ψ is applied logarithmic, k_f is applied exponential). This uncertainty provides the basis for the assessment of the water content (θ) as a relevant measurement parameter.

SYSTEMS

The comprehensive characterization of materials is prerequisite to understand processes in large (geo)-technical systems and their manipulation. This can be achieved best by the knowledge of material properties, measuring methods to determine water content and processes that describe the interaction of matter and water. Examples for technical systems in that sense are e.g. sealing systems for landfills and subsurface storage of waste, monitoring of soil water content over large areas using power lines, or monitoring system for groundwater recharge in the unsaturated zone.

Monitoring System for Surface Sealings

Configuration of Sealing System and Monitoring Layer

For the monitoring of the volumetric water content θ_v, the TAUPE TDR-system described in chapter 1.2.3.2 was used. Specifications for a monitoring system for surface sealing systems on landfills (figure 1) defined from legislating body (BAM, Federal Institute for Materials Research and Testing) are

- Detection of increase over more than an order of magnitude in permeability of a mineral sealing or capillary barrier,
- Detection of local relative variations in volumetric water content of more than 5 %,
- Positioning information of 100 m², that means a circle with radius around 5.5 m around true position.

The sealing system installed at the landfill is build up as a capillary barrier as shown in figure 2.

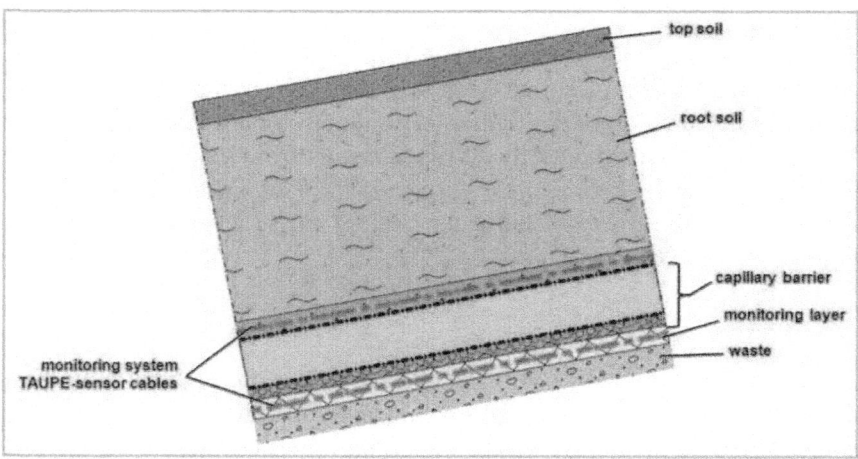

Figure 2. Simplified schematic of mineral capillary barrier consisting of a fine-grained material capillary layer above a coarse-grained material capillary block in a surface sealing system of a landfill.

Sensors of the monitoring system were installed in monitoring layers above and below the capillary barrier to determine the quantity of permeated water through the barrier. Water content of a soil layer situated below the capillary barrier without appreciable hysteresis is directly related to its permeability, i.e. the unsaturated hydraulic conductivity. Landfill regulations determine saturated hydraulic conductivity as quality parameter of mineral sealing systems. A critical limit of $< 5 \times 10^{-9}$ m/sec can be deduced from experimental relation between volumetric water content and unsaturated hydraulic conductivity, which can reach values of a few orders of magnitude below that limit. Materials used in a monitoring layer must meet the following requirements

- Its hydraulic properties must be defined unambiguously,
- Inflow from layers above must lead to a significant change in volumetric water content,
- Uniquely defined monitoring parameters for saturated/unsaturated hydraulic conductivity correspond to definite volumetric water contents. Percolation through monitoring layer must be equivalent to percolation in total sealing system.

Adequate for a monitoring layer are sandy to silty materials with a range of saturated hydraulic conductivity of 10^{-5} to 10^{-6} m/sec. A decrease in unsaturated hydraulic conductivity should be at least four to five orders of magnitude and cover the range of 10^{-10} m/sec to 10^{-5} m/sec. The monitoring layer then can discharge a break-through of the sealing system without building up backwater. Sensors are installed in monitoring layers at depth of around 2 m and 2.3 m, respectively.

Calibration of a Material for the Monitoring Layer

An adequate material was tested in the laboratory using measurement equipment for the determination of complex dielectric permittivity. It consists of a vectorial network analyzer and a coaxial type probe cell (see inset in figure 2) in the same frequency range of 100 MHz to 1100 MHz as used with TDR method.

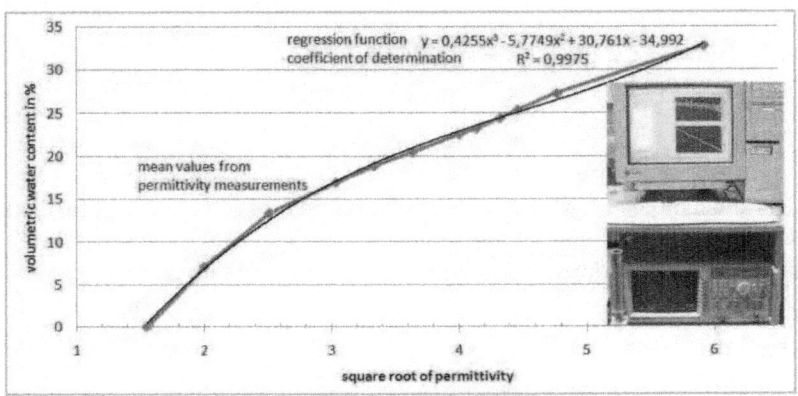

Figure 3. Material calibration function for the monitoring layer. Inset shows the permittivity measurement system with coaxial probe cylinder.

The material has been exposed to different amounts of water, with a part of it being dried at 105 °C in oven for evaluation of gravimetric water content related to dry mass. A second part was inserted in the coaxial probe cylinder for determination of permittivity. Figure 3 shows the relation between volumetric water content and square root of measured permittivity. Results from TDR measurement can directly be converted via the indicated regression function.

Monitoring System

A test site for a monitoring system for surface sealing has been established from 2004 to 2005 at the landfill situated in Oberweier/Germany (Figure 4). In two monitoring layers (see figure 1) 230 sensors have been installed and connected via 34 multiplexer to a TDR system.

To keep the length of the connecting coaxial cables between TDR system and sensors below 150 m, two central units with separate TDR devices cover 120 and 110 sensors, respectively. Sensor length is 10 m and distances between adjacent sensors are between 8 and 10 m, depending on hill slope. Data collection takes place two times a day.

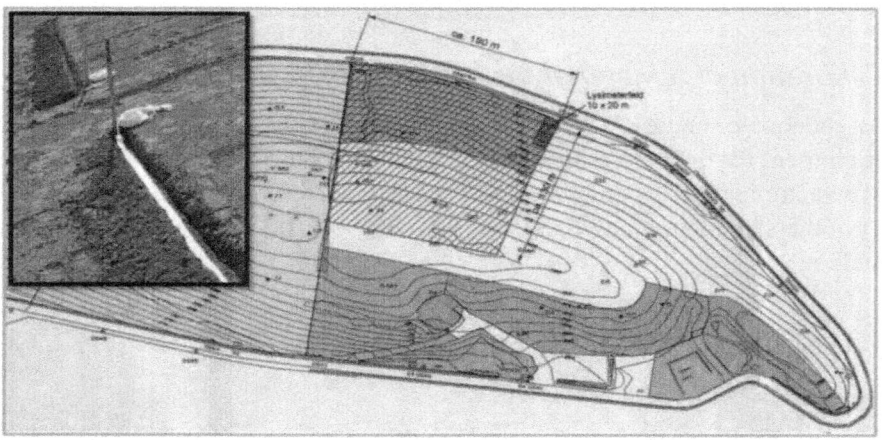

Figure 4. Plan of landfill in Oberweier. Sensors and measurement equipment has been installed in the shaded area of approximately 150 m x 130 m. Inset shows TAUPE TDR-cable sensors during installation in lower monitoring layer.

Data Evaluation

TDR signal data comprise only a part of the total signal length and is constraint to the transition between coaxial cable and start of sensor and shortly beyond end of sensor (see figure 5). First rise of the reflection signal occurs at start of sensor and second rise at end of sensor (Topp et al., 1980). Exact starting and ending points are defined by calculating the inflection points of the slopes to fit tangents to the curve and finding crossing points with horizontal lines. From time difference propagation time is calculated.

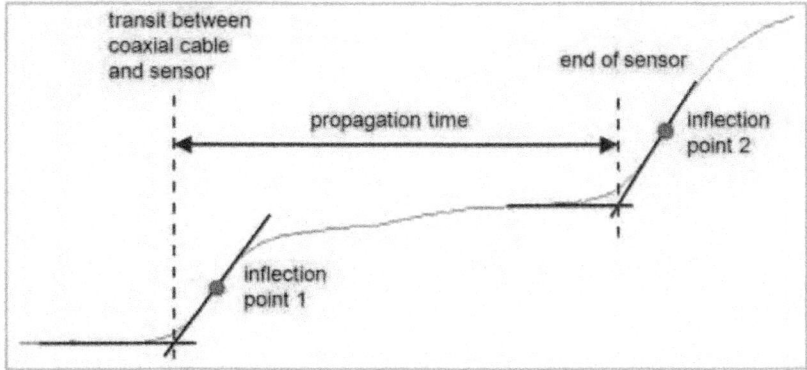

Figure 5. Typical form of TDR reflection signal and determination of propagation time.

Data from all sensors are automatically evaluated using an appropriate software system. Resulting propagation time defines an integral value for volumetric water content along a complete sensor according to the material calibration function in figure 2. Adding results for each sensor over time delivers variations in water content all over the landfill. To give an easier access to the hydraulic behavior at locations of different sensors the landfill is divided in vertical transects between top of the landfill and its base. This is shown for 2010 in figure 6 on eight sensors for both monitoring layers.

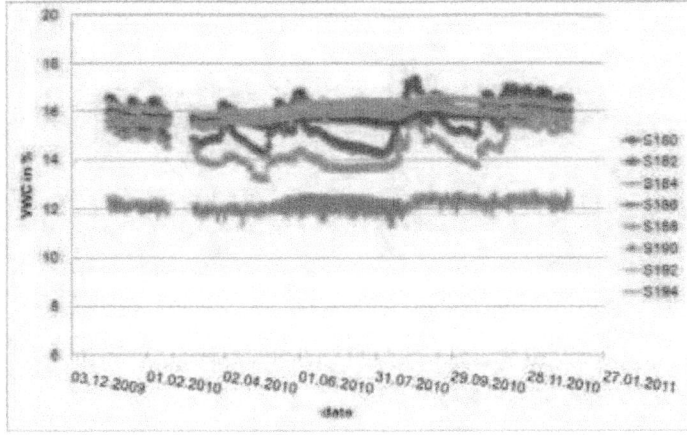

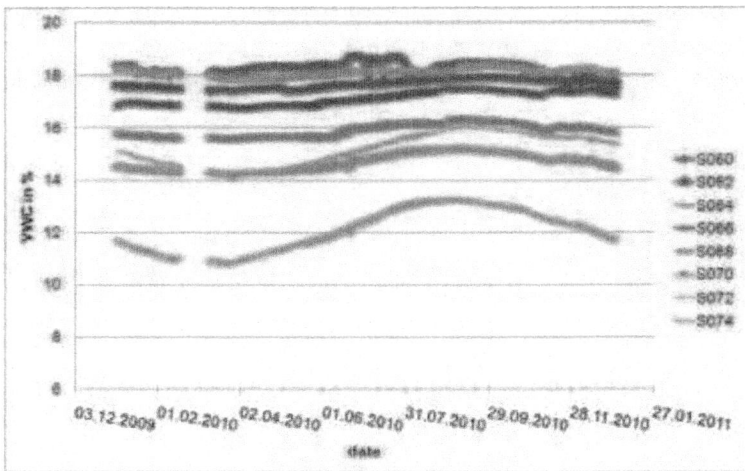

Figure 6. Volumetric water content as a function of time in monitoring layer above (left: sensors 180, 182, 184, 186 188, 190, 192, 194) and below (right: sensors 60, 62, 64, 66, 68, 70, 72, 74) capillary barrier. Data gap in March 2010 was due to a failure in power supply.

In the monitoring layer above capillary barrier discharge of precipitation at the surface can be found directly, especially at the borders of the landfill due to a poor connection to the area outside of the test site. Preferential flow parallel to the slope in layers with local secondary capillary barriers, which have been built unintentionally during installation process, leads to a more steady value of volumetric water content. Normally short time reactions describe the hydraulic system above the sealing.

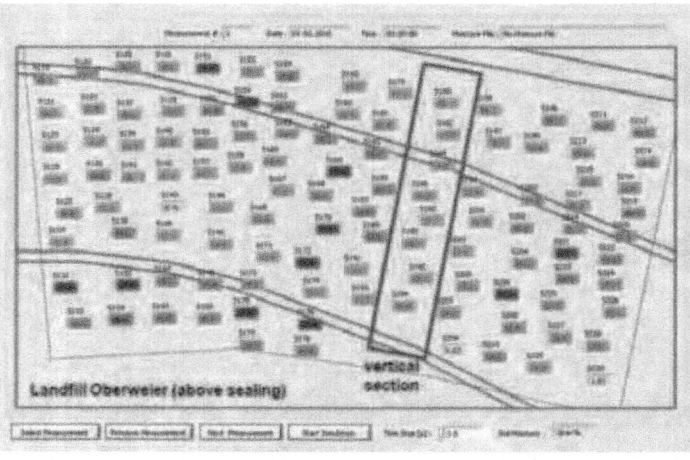

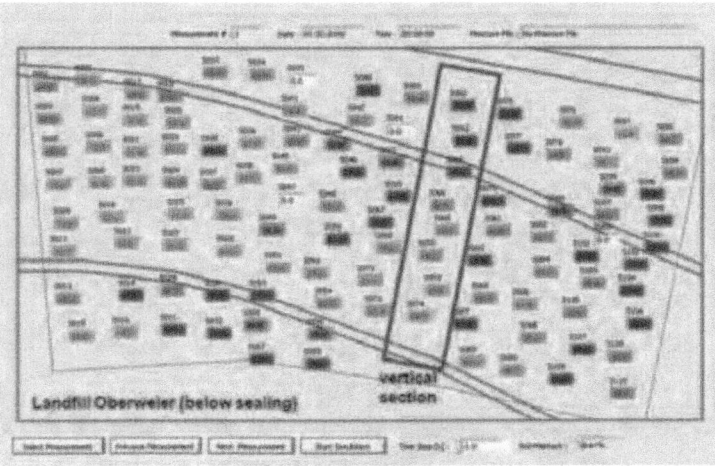

Figure 7. Intrinsic situation of volumetric water content in monitoring layer above (left) and below sealing system (right). Vertical sections according to figure 6.

In contrast, situation below the sealing system shows little short time variations, what demonstrates the functionality of the capillary barrier. Except at the borders little influences of discharge from the surface occur. Depending on chemical reactions in waste and seasonal temperature changes the volumetric water content can vary locally in the order of up to 2 % due to temperature depending permittivity of water.

Graphs in figure 7 show the situation above and below the capillary barrier at a certain time as colored graphs. Rectangular blocks show the volumetric water content of each sensor and the red color indicates possible problems due to locally high water content above arbitrarily chosen 18 %. Situation below the sealing shows more dark blue and red spots than above, which is a result of the installation since the first section below the sealing has been constructed during heavy rain in October 2004 and the second section with the capillary barrier was built in spring 2005 during the dry season. Water exchange with atmosphere via evapotranspiration is low due to depth of monitoring layers.

Hydraulic Conductivity

Volumetric water content of the layer below capillary sealing received from the monitoring system is the input parameter for the determination of hydraulic conductivity according to figure 8. In 2010 the sensors detected volumetric water contents between 12 and 18 %. Critical limit of 5×10^{-9} m/sec gives a monitoring value for volumetric water contents of 22 % and has not emerged during the observation period between 2005 and 2010.

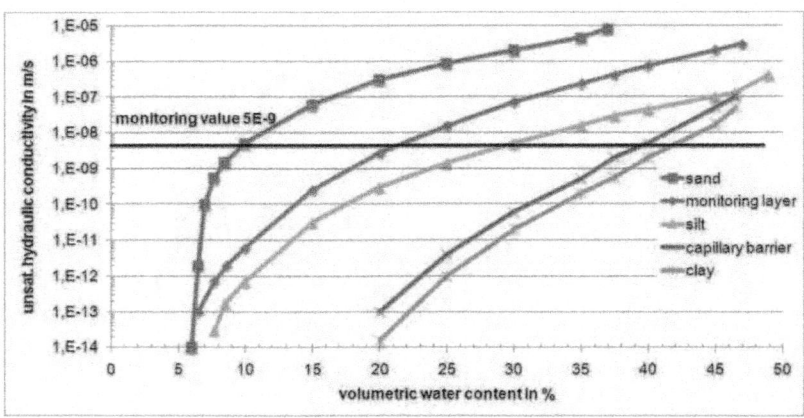

Figure 8. Hydraulic conductivity of different materials as function of VWC; material of monitoring layers fulfills requirements of BAM.

ACKNOWLEDGEMENT

We acknowledge support by Deutsche Forschungsgemeinschaft and Open Access Publishing Fund of Karlsruhe Institute of Technology.

REFERENCES

1.   Bachmann, J., Ellies, A., & Hartge, K. H. (2000). Development and application of a new sessile drop contact angle method to assess soil water repellency. Journal of Hydrology Vol. 231-232, pp. 66-75.

2.   Berger, K. (1998). Validierung und Anpassung des Simulationsmodells HELP zur Berechnung des WHH. BMBF-Schlussbericht, Berlin.

3.   Brandelik, A., & Hübner, C. (1999). Moisture monitoring with subsurface transmission line. Proceedings of Conference on Subsurface Sensors and Applications, Denver, June 1999.

4.   Brooks, R.H. and Corey, A.T. (1964). Hydraulic properties of porous media. Hydrological Papers, No. 3, Colorado State University.

5.   Brunauer, S., Emmet P.H., & Teller E. (1938). Adsorption of gases in multimolecular layers. J. Am. Chem. Soc., Vol. 60, pp. 309-319.

6.   Burdine, N.T. (1953). Relative permeability calculations from pore size distribution data. Petroleum. Transactions of the American Institute, of Mining, Metallurgical and Petroleum Engineers, Vol. 198, pp. 71-77.

7.   Butt, H.-J., Graf, K., & Kappl, M. (2006). Physics and Chemistry of Interfaces. Wiley-VCH Verlag, Weinheim.

8. Campbell, G.S. (1974). A simple method for determining unsaturated conductivity from moisture retention data. Soil Science, Vol. 117, No. 6, pp. 311-314.

9. Carman, P. C. (1937). Fluid flow through granular beds. Trans. Inst. Chem. Eng. Vol. 15, pp. 150-166.

10. Cox, R. G., 1986a. The dynamics of the spreading of liquids on a solid surface. Part 1. Viscous flow. Journal of Fluid Mechanics Vol. 168, pp. 169-194.

11. Cox, R. G., 1986b. The dynamics of the spreading of liquids on a solid surface. Part 2. Surfactants. Journal of Fluid Mechanics, Vol. 168, pp. 195-220.

12. Czurda, A. (1994). Multimineralische Abdichtungen. Schriftenreihe Angewandte Geologie Karlsruhe, Vol. 30, pp. 1-21.

13. Dang-Vu, T. & Hupka, J. (2005). Characterization of porous materials by capillary rise method. Physicochemical Problems of Mineral Processing, Vol. 39, pp. 47-65.

14. Dimitrov, A. S., Kralchevsky, P. A., Nikolov, A. D., Noshi, H., & Matsumoto, M. (1991). Contact angle measurements with sessile drops and bubbles. Journal of Colloid Interface Science, Vol. 145, pp. 279-282.

15. Dohrmann, R. & Kaufhold, S. (2010). Determination of exchangeable calcium of calcareous and gypsiferous bentonites. Clays and Clay Minerals, Vol. 58 , pp. 79-88.

16. Emmerich, K., Kemper, G., Königer, F., Schlaeger, S., Gruner, M., Hofmann, M., Nüesch, R.,& Schuhmann, R. (2007). HTV-1: A semi technical scale testing of a multi-layer hydraulic shaft sealing system. Proceedings of 3rd International Meeting on Clays in natural and engineered barriers for radioactive waste confinement, Lille, Frankreich, 2007

17. Emmerich, K., Kemper, G., Königer, F., Buqezi-Ahmeti, D., Gruner, M., Gaßner, W., Hofmann, M., & Schuhmann, R. (2008). Sandwich - Sealing system with equipotential layers for underground storage of hazardous waste to ensure homogeneous wetting of sealing layers and to enhance long term stability. Proceedings of Bodenkundliches Kolloquium des Instituts für Bodenkunde und Standortslehre der Uinversität Hohenheim, Stuttgart, Deutschland, 2008

18. Emmerich, K. (2011): Thermal analysis for characterization and processing of industrial minerals. In: EMU notes in mineralogy, Christidis, G., Vol. 9: Industrial mineralogy, Mineralogical Society in press.

19. Genuchten, M.Th. van (1980). A closed-form equation for predicting the hydraulic conductivity of unsaturated soils. Soil Sci. Soc. Am. J., Vol. 44, pp. 892-898.

20. Grace, H. P. (1953). Resistance and compressibility of filter cakes. Part I. Chem. Eng. Prog., Vol. 49, pp. 303-318.

21. Haude, W. (1958). Über die Verwendung verschiedener Klimafaktoren zur Berechnung der potentiellen Evaporation und Evapotranspiration. Meteorologische Rundschau, Vol. 11, pp. 96-99.

22. Hillel, D. (1980). Fundamentals of soil physics, Academic press, Orlando.

23. Huebner, C. (1999). Entwicklung hochfrequenter Messverfahren zur Boden- und Schneefeuchtebestimmung, Wissenschaftliche Berichte FZKA 6329, Karlsruhe.

24. Jentsch, G. (1992). Bilanzierung des Stoff- und Schadstoffeintrags in das Grundwasser unter besonderer Berücksichtigung der ungesättigten Zone. Schriftenreihe Angewandte Geologie, Vol. 17.

25. Jones, G. & Frizzell, L. D. (1940). A theoretical and experimental analysis of the capillary rise method for measuring the surface tension of solutions of electrolytes. Journal of Chemical Physics, Vol. 8, pp. 986 - 997.

26. Jones, G. & Ray, W. A. (1937). The surface tension of solutions of electrolytes as a function of the concentration. 1. A differential method for measuring relative surface tension. Journal of the American Chemical Society, Vol. 59, pp. 187-198.

27. Kahr, G. & Madsen, F.T. (1995). Determination of the cation exchange capacity and the surface area of bentonite, illite and kaolinite by methylene blue adsorption. Applied Clay Science, Vol. 9, pp. 327-336.

28. Kallioras, A., Piepenbrink, M., Schueth, C., Pfletschinger, H., Dietrich, P., Königer, F., & Rausch, R. (2010). Quantification of groundwater recharge through application of pilot techniques in the unsaturated zone. Proceedings of EGU General Assembly, Wien, Österreich, May, 2010.

29. Kaufhold, S. & Dohrmann, R. (2008). Comparison of the traditional Enslin-Neff method and the modified Dieng method for measuring the water uptake capacity. Clays and Clay Minerals, Vol. 56, pp. 686–692.

30. Kaufhold, S., Dohrmann, R., Klinkenberg, M. (2010). Water uptake capacity of bentonites. Clays and Clay Minerals, Vol.58, pp. 37-43.

31. Koeniger, F., Emmerich, K., Kemper, G., Gruner, M., Gaßner, W., Nüesch, R., & Schuhmann, R. (2008). Moisture spreading in a multi-layer hydraulic sealing system (HTV-1). Engineering Geology, Vol. 98, pp. 41-49

32. Koeniger, F., Emmerich, K., Kemper, G., Gruner, M., Gaßner, W., Stacheder, M., & Schuhmann, R. (2009). Monitoring of moisture spreading in a multi-layer hydraulic sealing system during saturation with a rock salt brine by TDR sensors. Proceedings of ISEMA, Helsinki, June, 2009.

33. Koeniger, F., Schmitt, G., Schuhmann, R., & Kottmeier, C. (2010). Free Line Sensing›, a new method for soil moisture measurements using high-voltage power lines. Near Surface Geophysics, Vol. 8, pp. 151-161

34. Konert, M., Vandenberge, J. (1997) Comparison of laser grain size analysis with pipette and sieve analysis: a solution for the underestimation of the clay fraction. Sedimentology, Vol. 44, pp. 523-535.

35. Kozeny, J. (1927). Soil permeability. Sitzungsber. Oesterr. Akad. Wiss. Wien. Math. Naturwisss. Kl. Abt. Vol. 136, pp. 271.

36. Latief, O. (2010): Korngrößenbestimmung an Tonmineralen Vergleich von Sedigraph-, Laserstreuung- und Zetasizermessungen, Proceedings of Annual Meeting German Ceramic Society, Hermersdorf, March 2010.

37. Lavi, B., Marmur, A., & Bachmann, J. (2008). Porous media characterization by the two-liquid method: Effect of dynamic contact angle and inertia. Langmuir, Vol. 24, pp. 1918-1923.

38. Madsen F.T., & Kahr G. (1992). Wasserdampfadsorption und spezifische Oberfläche von Tonen, DTTG-Tagungsband, Hannover, 1992.

39. Maubeuge, K. v. (2002). Investigation of bentonite requirements for geosynthetic clay barriers. Proceedings of Clay Synthetic Barriers Symposium, Nuremberg, April, 2002.

40. Maubeuge, K. v. & Egloffstein, T. (2004). Quality requirements for Bentonite in geosynthetic clay liners in the validity of test methods, In: Advances in geosynthetic clay liner technology: 2nd Symposium, Mackey, R. E. & Maubeuge K.v., pp. 11-30, ASTM International, ISBN 0-8031-3484-3, Mayfield.

41. Moore, D. & R.C. Reynolds, Jr. (1997). X-Ray Diffraction and the Identification and Analysis of Clay Minerals, 2nd ed.: Oxford University Press, New York

42. Mualem, Y., (1976a). A new model for predicting the hydraulic conductivity of unsaturated porous media. Water Resour. Res., Vol. 12, No. 3, pp. 513-522.

43. Mullins, C.E. (1991). Matric potential, In: Soil analysis - Physical methods, Smith, K. A., & Mullins, C.E., pp. 75-110, Marcel Dekker Inc., 0-8247-8361-1, New York.

44.   Neff, H. K. (1959). Über die Messung der Wasseraufnahme ungleichförmiger bindiger anorganischer Bodenarten in einer neuen Ausführung des Enslingerätes. Bautechnik, Vol. 39, No. 11, pp. 415-421.

45.   Omotoso, O., McCarty, D., Hillier, S. & Kleeberg, R. (2006). .... Clays and Clay Minerals, Vol. 54, No. 6, pp. 748–760.

46.   Passioura, J.B. (1980): The Meaning of Matric Potential. Journal of Experimental Botany, Vol. 31, No.123, pp. 1161-1169.

47.   Petersen, L. W., Moldrup, P., Jacobsen, O. H., & Rolston, D. E. (1996). Relations between specific surface area and soil physical and chemical properties. Soil Sci., Vol. 161, No. 1, pp. 9-21.

48.   Philip, J.R., Knight, J.H. (1974). On solving the unsaturated flow equation: 3. new quasianalytical technique, Soil Science, Vol. 117, No.1, pp. 1-13.

49.   Pusch, R. (2001). The Buffer and Backfill Handbook, Part2: Materials and Techniques. Technical Report TR-02-12, Swedish Nuclear Fuel and Waste Management Co, ISSN 1404-0344, Stockholm.

50.   Pusch, R. (2002). The Buffer and Backfill Handbook, Part1: Definitions, basic relationships, and laboratory methods. Technical Report TR-02-20, Swedish Nuclear Fuel and Waste Management Co, ISSN 1404-0344, Stockholm.

51.   Ramirez-Flores, J. C., Bachmann, J., & Marmur, A. (2009). Direct determination of contact angles of model soils in comparison with wettability characterization by capillary rise. Journal of Hydrology, Vol. 382, pp. 10-19.

52.   Scheffer, F. (1992). Lehrbuch der Bodenkunde (13. Ed.), Enke, ISBN 3-432-84773-4, Stuttgart

53.   Schlaeger, S. (2005). A fast TDR-inversion technique for the reconstruction of spatial soilmoisture content . Hydrology and Earth System Sciences, Vol. 9, pp. 481–492.

54.   Schmugge, T. J., Jackson, T. J., & McKim, H. L. (1980). Survey of methods for soil moisture determination. Water Resour. Res., Vol. 16, No. 6, pp. 961-979.

55.   Schultze, B., Zurmuehl, T., Durner, W. (1996). Untersuchung der Hysterese hydraulischer Funktionen von Böden mittels inverser Simulation, Mitteilungen der deutschen bodenkundlichen Gesellschaft, Vol. 80, pp. 319-322.

56.   Schuhmann, R. (2002). Kontrolle von Barrieren: Bestimmung der hydraulischen Leitfähigkeit an Hand des Bodenwassergehaltes.

Mitteilungen des Instituts für Wasserwirtschaft und Kulturtechnik der Universität Karlsruhe, Vol. 219.

57. Siebold, A., Walliser, A., Nardin, M., Oppliger, M., & Schultz, J. (1997). Capillary Rise for Thermodynamic Characterization of Solid Particle Surface. Journal of Colloid and Interface Science, Vol. 186, pp. 60-70.

58. Topp, G.C., Davis, J. L., & Annan, A.P. (1980). Electromagnetic Determination of Soil Water Content: Measurements in Coaxial Transmission Lines. Water Resourc. Res., Vol. 16, No. 3, pp. 574-582.

59. Voinov, O. V. (1976). Hydrodynamics of wetting. Fluid Mechanics, Vol. 11, pp. 714-721.

60. Yang, Y. W. & Zografi, G. (1986). Use of the Washburn-Rideal equation for studying capillary flow in porous media. Journal of Pharmaceutical Sciences, Vol. 75, pp. 719-721.

Chapter 3

GEOTECHNICAL, MINERALOGICAL AND CHEMICAL CHARACTERIZATION OF THE MISSOLE II CLAYEY MATERIALS OF DOUALA SUB-BASIN (CAMEROON) FOR CONSTRUCTION MATERIALS

Elisabeth Olivia Logmo[1], Gilbert François Ngon Ngon[1], Williams Samba[1], Michel Bertrand Mbog[2], and Jacques Etame[1]

[1]Laboratory of Subsurface, Department of Earth Sciences, Faculty of Science, University of Douala, Douala, Cameroon

[2]Laboratory of Environmental Geology, Department of Earth Sciences, Faculty of Science, University of Dschang, Dschang, Cameroon

ABSTRACT

Geotechnical tests conducted on clayey materials of Missole II, Douala sub-basin of Cameroon showed that these materials present: fines particles (55 to 78 wt.%), sand (22 to 44 wt.%), and plasticity index of 13.8 to 21.6%. The X-ray diffraction (XRD) and the chemical analysis revealed a kaolinite amount of 46 to 56 wt.%, 19 to 27 wt.% of illite, 12 to 19 wt.% of quartz, 3 to 5 wt.% of goethite, 2 to 5 wt.% of hematite, 1.5 to 5 wt.% of anatase, 2 to 3 wt.% of feldspar-K with 52.87 to 63.11 wt.% of SiO_2, 18.08 to 24.31 wt.% of Al_2O_3, 3.28 to 11.45 wt.% of Fe_2O_3 and a small content of bases (<2 wt.%). The results of geotechnical tests combined to those of the XRD and the chemical analysis showed that the Missole II clayey materials are suitable for the manufacture of bricks, tiles and sandstones.

INTRODUCTION

Clay-rich materials are intensively used in the manufacturing of ceramics and as construction materials. However, the clayey deposit types are known as sedimentary, alluvial, and residual. A good knowledge of their occurrences, quantity and properties is required for their efficient exploitation.

In the Douala sedimentary sub-basin (South-Cameroon, Central Africa), some studies are done concerning the stratigraphic and tectonic evolution including [1-13]. Others studies are done to set up industrial units for manufacturing construction materials and ceramics [14] or concerning the mineralogical and chemical or thermal characteristics of the clay sediments [15-21].

In the Missole II area (Douala sub-basin), a geological study is carried out to locate and describe the clayey material outcrops and their provenance [22-24]. Also, some authors showed the possibility to obtain good ceramic building materials by mixing silica, feldspars, and kaolinitic and illitic clay of the Missole II area [25]. However, despite the preliminary works done on the field to describe clayey materials and to determine their sedimentation evolution, no further physical, mineralogical and chemical study is carried out to show their characteristics in order to their applications.

To that effect, the objective of this study is to associate the geotechnical characteristic to the mineralogical and chemical compositions of the clay occurrences of the Missole II deposit in order to evaluate its suitability for manufacturing of construction materials and ceramics.

GEOGRAPHIC AND GEOLOGICAL SETTING

Missole II is located on the Eastern part of the Douala sub-basin (Cameroon, Central Africa) between latitude 3°59› - 3°54›N and longitude 9°54› - 9°58›E. It is located within a humid equatorial climatic zone. Annual rainfall ranges between 3000 and 5000 mm, and the annual average temperature is 26°C. The vegetation is a dense rainforest transformed by human activities [26]. The geomorphology of the study area is a domain of the Cameroon coastal plain with low altitudes (40 - 120 m). The Missole II area shows hills with flat and sharp summits and is deeply dissected by V and U shaped valleys of MBongo, Bongougou, Missolo and Bongo the main rivers of the area. According to the geological map of SNH/UD report [12], the relative age of the Missole II sediments is Paleocene-Eocene corresponding to the N'Kapa Formation **Figure 1**.

The lithostratigraphy of Douala sub-basin is made of seven major Formations related to its geodynamic and sedimentary evolution [10-12]. 1) The synrift period represented by the Mundeck Formation (Aptian-Cenomanian) is discordant onto the Precambrian basement and consists of continental and fluvio-deltaic deposits, i.e., clays, coarse-grained sandstones, conglomerates. The postrift sequence includes; 2) the Logbadjeck Formation (Cenomanian-Campanian), discordant onto the Mundeck Formation and composed of micro conglomerates, sand, sandstone, limestone, and clay; 3) the

Logbaba Formation (Maastrichtian), mainly composed of sandstone, sand and fossiliferous clay; 4) the N'kapa Formation (Paleocene-Eocene), rich in marl and clay with lenses of sand and fine to coarse-grained crumbly sandstone; 5) the Souellaba Formation (Oligocene) lying unconformably on N'kapa deposits and characterized by marl deposits with some interstratified lenses and sand channels; 6) the Matanda Formation (Miocene), dominated by deltaic facies interstratified with volcanoclasties layers; and 7) the Wouri Formation (Plio-Pleistocene) which consists of gravelly and sandy deposits with a clayey or kaolinic matrix.

RAW MATERIALS AND EXPERIMENTAL METHODS

The raw material used in this study comes from four representative clayey profiles of the Missole II area with three profiles of the interfluves along the Douala-Edéa road and one from the pit drilling on the lower slope of the valley. A geological survey shows different types of sediments with micro conglomerates, sandstones, fragments of ferruginous duricrusts and clays. Clayey layers are overlain upwards by sandstones or micro conglomerates, ferruginous duricrusts and sandy-clays, and occupy the lower part of the profiles. Four clayey facies identified with different mixed textures like sandy-clay, clayey-silt and silty-clay are mainly of sedimentary origin [23, 24].

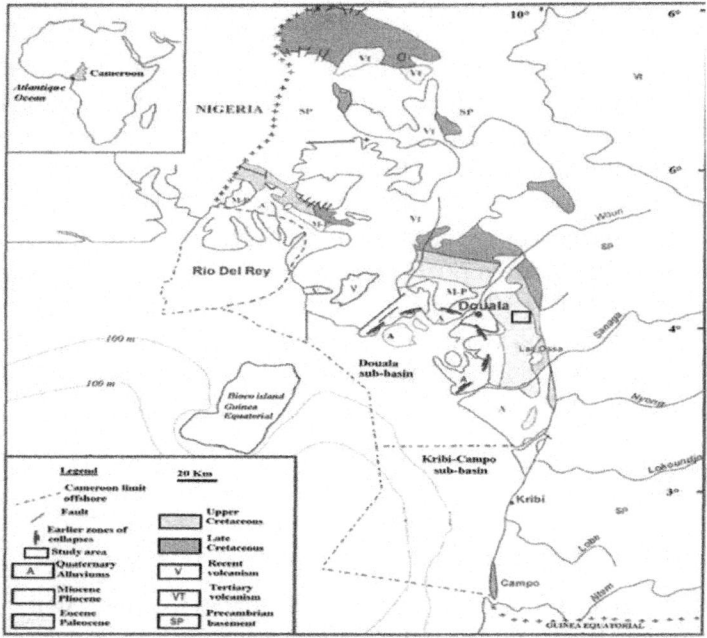

Figure 1. Geological sketch map of Cameroonian coastal basins (SNH/UD, 2005).

The average thickness of the exploitable layers is 2.5 m. Two clayey samples collected from clay layer of each representative profile for mineralogical and chemical data are mixed to obtain average sample which served to realize the geotechnical analyses. A sufficient quantity of the single mixture of 2 to 3 kg of sediments is collected from a meter-long groove. The sample is analyzed for physical, mineralogy and chemistry.

The particle size distribution has been achieved in two steps: 1) a conventional sieving for the 63 to 2000 µm fractions; 2) using a sedigraph 5000 in automatic procedure for clay and silt fractions. The liquid limit is measured by the method of the dish of Casagrande (wL) and the plastic limit by the method of the roller (wP). The blue methylene value (Vb) is determined on the total sample.

One hundred grams of each homogenized sample is grounded to −200 mesh (0.075 mm) in an agate mortar for chemical and XRD mineralogical study. Mineral identification is performed using a Setsys 2400 apparatus from SETARAM 85 equipped with a DSC 1500 heat system with Pt crucibles for thermal analysis, from room temperature up to 1100°C using a rate treatment of 10°C·min^{-1}and alumina heat treated at 1500°C serving as reference material. For XRD, a Brünker diffractometer D8 ADVANCE with a copper source (λ = 1.5489 Å) is used on bulk and fine (<2 µm) samples, working under 40 kV and 40 mA. The exposure time for qualitative analysis is 2 h. Mineralogical phases are identified (JCPDS, 1998). Semi-quantitative analysis is performed [27]. For microscopic analysis, clay samples are examined with a scanning electron microscope (SEM) (Cambridge stereos can 200) coupled with an energy dispersive spectra microprobe (EDS). Homogenized powder of sediment sample is chemically analyzed for major and trace elements by ICP-AES after dissolution using acid digestion procedure with HF, HNO_3, and $HClO_4$. Classification of the clayey materials is performed by Autret (1983) method [28,29]. This method differentiates lateritic materials to non lateritic materials. It is based on the ratio S/R where, S is the ratio between SiO_2 concentration and the molar mass and R the ratio between Al_2O_3 plus Fe_2O_3concentrations and their molar mass. In fact, for true lateritic material S/R is less than 1.33, for lateritic rocks it is 1.33 to 2, and for clay material it is more than 2.

RESULTS

Geotechnical Characteristics

The particle size distribution of the raw basic materials shows that the samples consist of 23 to 45 wt.% of sand, 17 to 33 wt.% of silts and 34 to 45 wt.% of clayey fractions. The geotechnical characteristics of the samples are reported in **Table 1**. These results are permitted to deduce that the class of these samples is in fine fractions [30]. The plasticity index (13.8% to 21.6%) is varied between 12 and 25, and shows that these materials are averagely plastic [31]. This plasticity is presumably a consequence of the average content of clayey minerals and quartz. The methylene blue value (0.4 to 0.91 g/100g) of the concerned materials, which is corroborated with the plasticity index, may suggest the absence of swelling clay minerals.

Mineralogical Composition

DTA analysis

Figure 2 represents the DTA analysis curves obtained for the raw materials M2P3 and M2A3 at room temperature up to 1100°C using a rate treatment of 10°C·min⁻¹. Differential scanning calorimetric (DSC) observed on these DTA analysis curves shows endothermic phenomenon towards 110°C associated to the physisorbed water and another endothermic phenomenon towards 320°C due to the dehydroxylation of goethite.

Table 1. Geotechnical characteristics of the samples.

Geotechnical characteristics	Results			
Average samples	M2A2	M2P3	M2P4	M2A3
Depth (cm)	600 - 750	700 - 800	700 - 800	50 - 150
Particles size distribution (wt.%)				
Sand (2000 - 63 µm)	45	31.3	44	22
Silts (63 - 2 µm)	21	25.5	17	33
Clays (<2 µm)	34	43.2	39	45
Atterberg limits				
Liquid limit, wL (%)	35.11	40.87	32.9	47.45
Plasticity limit, wP (%)	19.84	22.92	19.07	25.81
Plasticity index, PI (%)	15.27	17.95	13.83	21.64
Blue methylene value of the total sample (g/100g)	0.4	0.82	0.7	0.91

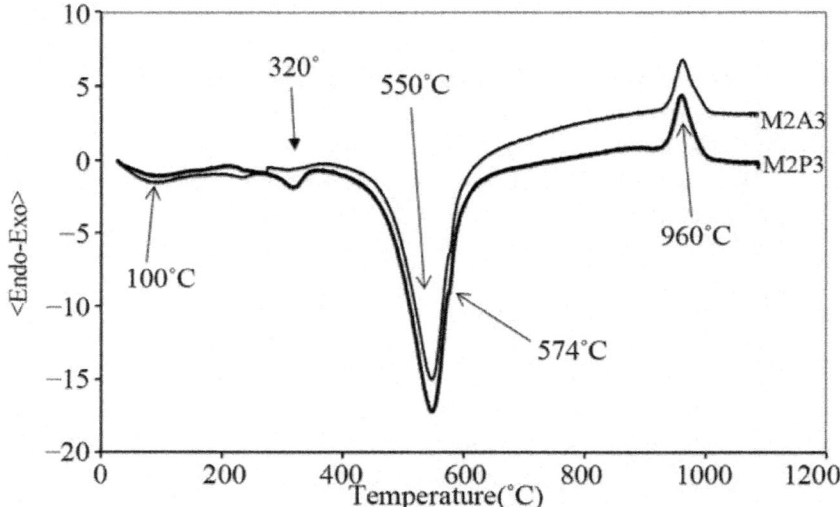

Figure 2. DTA curves of M2A3 and M2P3 materials for a heating rate of 10°C/min.

A specific endothermic transformation at 400°C - 600°C is due to the loss of structural hydroxyl groups of the kaolinite and an exothermic peak at 900°C - 1000°C due to the structural reorganization of the metakaolinite [32,33].

X-Ray Diffraction

The diffractograms of powder of the bulk samples ground until <80 μm revealed that clayey materials are kaolinitic and illitic and have permitted the detection of the following mineral phases: kaolinite, quartz, illite, goethite, hematite and accessory feldspar-K and anatase [24]. These mineral phases were usually present in theclayey materials of the Douala sub-basin [16,17,21].

The diffractograms of <2 μm fraction of **Figure 3** also indicated the presence of the same mineral phases observed in the total samples with kaolinite, illite, quartz, goethite, hematite and accessory feldspar-K and anatase. Comparison of the minerals detected, these mineral phases are present in the fraction of the smaller and thicker granulometry.

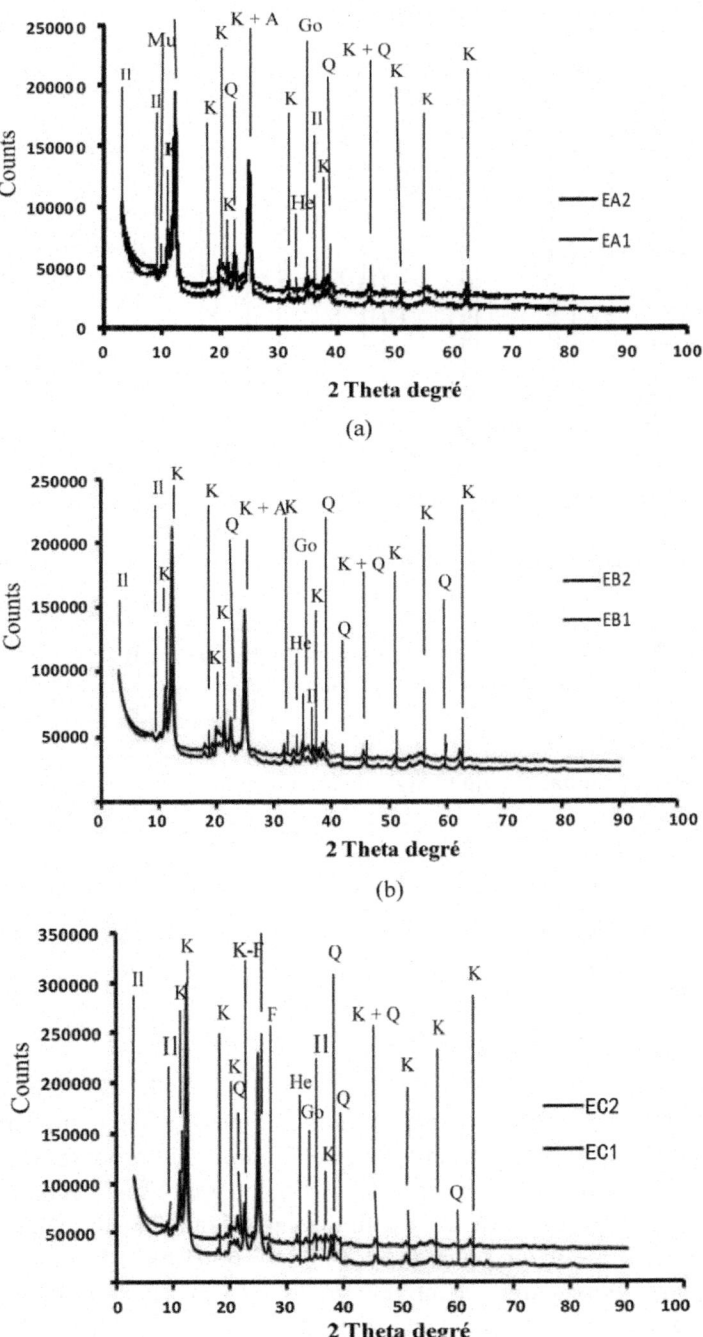

(a)

(b)

(c)

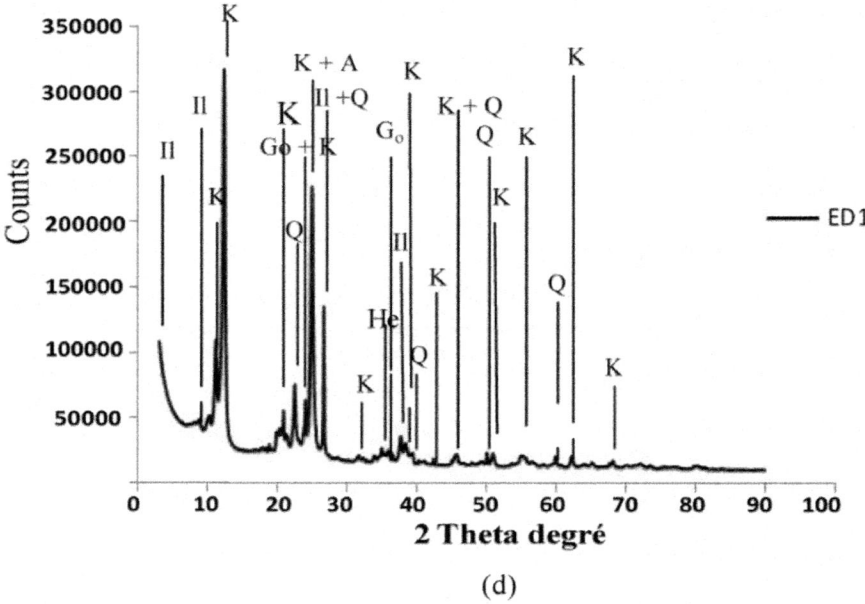

(d)

Figure 3. XRD patterns of <2 μm clay samples from Missolle II area of (a), (b), (c) and (d) profiles. A: Anatase; G: Goethite; He: Hematite; Il: Illite; K: Kaolinite; Q: Quartz; F: K-feldspar.

Table 2. Semi-quantitative mineralogical composition of the samples (wt.%).

Profiles										Profiles of the pit	
		Profiles of the interfluves									
Profiles		a			b			c			d
Samples	EA1	EA2	M2A2	EB1	EB2	M2P3	EC1	EC2	M2P4	ED1	M2A3
Depth (cm)	600 - 700	700 - 750	600 - 750	700 - 750	700 - 750	750 - 800	700 - 750	750 - 800	700 - 800	50 - 100	50 - 150
Kaolinite	53.9	52.1	53.6	46.4	50.4	48.1	47.7	48.5	48.5	47.8	48.4
Illite	23.5	20.5	20.8	26.3	19.9	23.4	26.6	19.2	20.8	25.0	23.6
Quartz	16.1	15.0	17.1	18.8	17.7	18.2	12.5	17	16.9	13.7	16.1
Goethite	3.0	4.0	4.2	3.9	3.0	4.0	3.2	4.4	3.8	5.0	4.1
Hematite	-	2.2	1.6	-	3.0	2.1	3.2	4.4	3.6	2.9	3.2
Anatase	3.5	3.6	3.3	4.6	-	3.3	4.3	1.5	3.3	2.9	3.1
Feldspath-K	-	-	-	-	3.0	1.2	-	2.2	1.6	-	-

The Table 2 gives the semi-quantitative mineralogical composition of the mineral phases of the samples. It reported an important proportion of kaolinite (46 to 56 wt.%), appreciable amounts of illite (19 to 27 wt.%) and quartz (12 to 19 wt.%), which are in relation with the value of the plasticity index (13.8 to 21.6%) and the methylene blue value (0.8 to 1.6 g/100g). The amounts of the other mineral phases are respectively goethite (3 to 5 wt.%), hematite (2 to 5 wt.%), anatase (1.5 to 5 wt.%) and feldspar-K (2 to 3 wt.%).

Microscopic Analysis

Figure 4 shows a SEM observation performed on samples of grey and mottled, with sandy-clay, silty-clay and clayey-silt texture. In clayey materials, piles of disordered kaolinites are observed with various sizes and irregular forms, which indicated that kaolinites of sedimentary clayey material of the Missole II area are poorly crystallized [34].

Chemical Composition

The chemical composition of the samples is given on **Table 3**. The results of the chemical analysis shows the presence of important amounts of SiO_2 (52.87 to 63.11 wt.%), associated with an appreciable amounts of Al_2O_3 (18.08 to 24.31 wt.%) and Fe_2O_3 (3.28 to 11.45 wt.%) and a small content of bases (<2 wt.%, $CaO + MgO + Na_2O + K_2O$).

These results show that quartz, alumino-silicates compounds and iron minerals are predominate in the samples.

M2P3

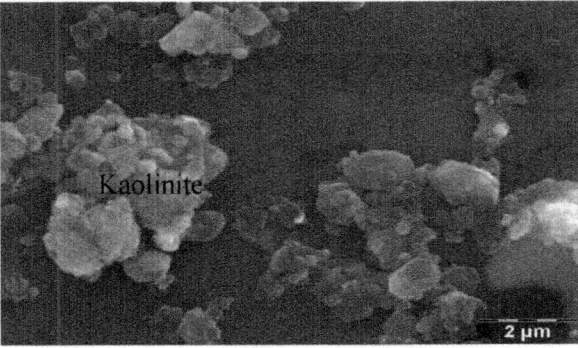

M2A3

Figure 4. SEM images of M2P3 and M2A3 clayey materials of Missole II.

Table 3. Chemical composition of the sample.

				Profiles of the interfluves							Profiles of the pit	
Profiles			a			b			c			d
Samples	DL	EA1	EA2	M2A2	EB1	EB2	M2P3	EC1	EC2	M2P4	ED1	M2A3
Depth (cm)		600 - 700	700 - 750	600 - 750	700 - 750	700 - 750	750 - 800	700 - 750	750 - 800	700 - 800	50 - 100	50 - 150
SiO_2	0.01	63.11	56.85	58.01	54.97	57.24	56.23	52.87	54.02	54.08	63.1	62.84
Al_2O_3	0.01	18.94	24.31	23.06	22.16	19.1	20.87	23.24	20.59	21.01	18.08	18.61
Fe_2O_3	0.01	3.28	3.45	3.39	5.96	7.54	8.02	3.51	11.45	8.89	4.0	3.54
CaO	0.01	0.04	0.02	0.02	0.02	0.2	0.02	0.02	0.02	0.02	0.02	0.02
MgO	0.01	0.1	0.1	0.1	0.13	0.08	0.1	0.13	0.08	0.12	0.1	0.1
Na_2O	0.01	0.06	0.06	0.06	0.1	0.03	0.1	0.06	0.06	0.06	0.06	0.06
K_2O	0.01	0.43	0.4	0.49	0.77	0.21	0.3	0.47	0.47	0.47	0.6	0.7
TiO_2	0.01	1.02	1.38	1.42	0.99	1.1	1.0	1.2	1.2	1.2	0.78	0.83
MnO	0.01	0.01	0.01	0.01	0.01	0.01	0.01	0.01	0.01	0.01	0.01	0.01
P_2O_5	0.01	0.02	0.04	0.04	0.04	0.04	0.03	0.02	0.05	0.05	0.03	0.03
Cr_2O_3	0.01	0.01	0.01	0.01	0.01	0.02	0.01	0.01	0.02	0.01	0.01	0.01
LOI		13.02	13.12	13.29	14.21	13.85	13.21	17.38	11.81	14.01	14.4	12.34
Total		99.94	99.75	99.9	99.37	99.22	99.9	99.29	99.79	99.93	99.25	99.34

DL: Detection limit; LOI: Loss on ignition.

The weak content of oxide calcium or potassium has proven that the calcitic and potassic minerals like calcite or feldspar are absent or are in a weak proportion. With S/R more than 2, the analyzed samples are clays [28].

DISCUSSION

The particle size in this study revealed texture classes: silty-clay, clayey-silt and sandy-clay described in many deposits of the Douala sub-basin [17,21]. These classes characterized very sticky and plastic materials in accordance with some field characteristics such as smoothness, stickiness and plasticity [35]. Indeed, the fine particles and especially the very fine clay minerals determine the plasticity of an earthly material [32]. The plasticity index (13.8% to 21.6%) of the above clayey samples is supported by the high amounts of fine particles (silt + clay) and fine clay minerals. The amount of clay fraction is used to determine the manufactured products as is shown in **Table 4**[36]. However, particle size and plasticity index are amongst the main geotechnical characteristics used to determine a convenient choice of construction material. In this case, taking into account of the amounts of sand (22 to 44 wt.%), fine particles notably the clay fractions (34 to 45 wt.%), and plasticity index, the Missole II clayey materials are convenient to the manufacturing of bricks, tiles and sandstones [37-39].

The mineralogical composition and geotechnical characteristics of a construction material must be known simultaneously in order to make a suitable choice for this raw basic material. It is well known that clay minerals have detrimental effects on the geotechnical properties; since some swelling clay minerals such as smectites absorb more water than the other (e.g. kaolinite). The volumetric shrinkage which increases with swelling clay minerals such as smectites constitutes an important parameter to take into account for civil public engineering [29,40]. The amount of clay minerals and especially swelling clay minerals contribute to the increasing of volumetric shrinkage and the swelling phenomena. Some authors joined the swelling potential and/ or volumetric shrinkage to two parameters [36]. A first ordering is based on the plasticity index and the percentage of fine clayey fraction (<2 μm size).

Table 4. Relation between particles size and type of ceramic product to realize (after Cere and Mazel, 1993).

$\phi < 2$ μm (wt.%)	Ceramic products to realize
5 - 25	Full bricks
25 - 35	Perforated bricks
35 - 45	Hollow bricks, drains
40 - 50	Tiles, hordes

A second method proposes to use the liquid limit and the percentage of fine particles which size is less than 74 μm [41]. In this case, the studied materials, which have plasticity index between 12% and 25%, 34 and 45 wt.% of clayey fraction (<2 μm size), 54 and 78 wt% of fine fraction (<80 μm size) and liquid limit between 32% and 47% are very low to moderate swelling materials as is indicated in **Table 5**. This observation is corroborated with the mineralogical composition which shows that fine fraction is mainly composed of kaolinite and few illites, as well as smectites are absent.

The geotechnical and mechanical characteristics of clayey materials are essentially depending on their chemical and mineralogical compositions as well as of the minerals distribution [37,39]. In this study, the clayey samples mainly constituted of poorly crystallized kaolinite (46 to 56 wt.%), which is in agreement with the loss on ignition (12% to 17%), illite (19 to 27 wt.%) and quartz (12 to 19 wt.%), are adequate for the composition of ceramic pastry [42] and for the manufacturing of bricks and tiles [14,37]. Also, clayey materials are siliceous ($SiO_2 > 60$ wt.%), with their alumina (Al_2O_3) content less than 35 wt.% and appreciable amount of iron (<10 wt.% with the exception of EC2 sample) is suitable for the manufacturing of tiles and bricks [41]. However, the weakest amounts of bases <2 wt.% (oxides of calcium, magnesium, sodium and potassium) which are the smelting or cimentitious compound corroborated with the flexural strength less than 10 MPa of the Missole II clayey material [25].

CONCLUSIONS

The following conclusions can be drawn out from the present work:

1. The geotechnical, mineralogical and chemical results show that the clayey materials of the Missole II, Douala sub-basin of Cameroon are mainly constituted of sand (23 to 45 wt.%), silts (17 to 33 wt.%) and clays (34 to 45 wt.%) with plasticity index of 13.8% to 21.6%.

2. Kaolinite, illite and quartz are the main minerals.

3. Clayey materials are also siliceous and aluminous, and have appreciable amount of iron in most clayey samples (Fe_2O_3 less than 10 wt%).

Table 5. Swelling potential of soils according to their plasticity index, liquid limit and respectively their <2 and <74 μm fractions.

PI (%)	% <2 μm	% <74 μm	wL (%)	Swelling potential
>35	>95	>95	>60	Very high
22 - 35	60 - 95	60 - 95	40 - 60	High
18 - 22	30 - 60	30 - 60	30 - 40	Moderate
<18	<30	<30	<30	Low

4. Based on the obtained geotechnical, mineralogical and chemical results, and the literature data, the studied materials are suitable for the manufacturing of bricks, tiles and sandstones.

5. These clayey materials present poor mechanical characteristics (feeble flexural strength and great fragility) partly linked to the chemical and mineralogical nature fundamentally complex of natural clays.

ACKNOWLEDGEMENTS

The authors acknowledge the financial support of the University of Douala and the geochemical analysis of Geo Labs (Geoscience Laboratories) in Canada. Many thanks are also given to Professor David Smith, Director of the GEMH laboratory at ENSCI and Professor Gisèle Lecomte of Centre Européen de la Céramique, Limoges, France, and to all of their collaborators.

REFERENCES

1. Y. C. Belmonte, "Stratigraphie du bassin sédimentaire du Cameroun," Proceeding on 2nd West African Micropaleontology Colloquium, Ibadan, 1-5 September 1966, pp. 7–24.

2. M. E. Brownfield and R. R. Charpentier, "Geology and Total Petroleum Systems of the West-Central Coastal Province (7203) West Africa," USGS: Geological Survey Bulletin, 2006, 2207-B 52 p.

3. E. Dartevelle and P. Brebion, "Mollusques fossiles du Crétacé de la Côte occidentale d'Afrique du Cameroun à l'Angola,"Annales du Musée Royal Congo Belge, Sciences Géologiques de Tervuren I-Gastéropodes, Vol. 8, No. 20, 1956, pp. 1-128.

4. E. Dartevelle, S. Freinex and J. Sornay, "Mollusques fossiles du Crétacé de la Côte occidentale d'Afrique du Cameroun à l'Angola,"Annales du Musée Royal Congo Belge, Sciences Géologiques de Tervuren II-Lamellibranches, Vol. 8, No. 20, 1957, pp. 1-271.

5. J. F. Dumort, "Identification par la Telédétection de l'Accident de la Sanaga (Cameroun)," Géodynamique, Vol. 1, No. 3, 1968, pp. 13-19.

6. J. B. Meyers, B. R. Rosendahl and H. Groschel-Becker, "Deep Penetrating MCS Imaging of the Rift-to-Drift Transition Offshore Douala and North Gabon Basins West Africa," Marine Petrology Geology, Vol. 13, No. 7, 1996, pp. 791-835. doi:10.1016/0264-8172(96)00030-X

7. D. Reyre, "Histoire géologique du bassin de Douala," In: D. Reyre, Ed., Symposium sur les Bassins Sédimentaires du Littoral Africain, Association du Service Géologique d'Afrique, IUGS, 1966, pp. 143-161.

8. P. R. N. Ngaha, "Contribution à l'Etude Géologique, Stratigraphique et Structurale de la Bordure du Bassin Atlantique du Cameroun," Thèse 3e Cycle, Université de Yaoundé, 1984.

9. P. Maurizot, A. Abessolo, J. L. Feybesse, V. Johana and P. Lecomte, "Synthèse des Travaux de 1978 a` 1986," Rapport 85CM066, 1986.

10. J. M. Regnoult, "Synthèse Géologique du Cameroun," D.M.G., Yaoundé, Cameroun, 1986.

11. F. R. Nguene, S. Tamfu, J. P. Loule and C. Ngassa, "Paleoenvironments of the Douala and Kribi/Campo Subbasins in Cameroon, West African," Colloque de Géologie Africaine, Libreville, Recueil des Communications, Géologie Africaine, 6-8 May 1991, 1992, pp. 129-139.

12. SNH/UD, "Stratigraphie Séquentielle et Tectonique des Dépôts Mésozoïques Synrifts du Bassin de Kribi/Campo," Rapport Non Publié, 2005.

13. C. S. Manga, "Stratigraphy, Structure and Prospectivity of the Southern Onshore Douala Basin Cameroon—Central Africa," In: M. J. Ntamak-Nida, G. E. Ekodeck and M. Guiraud, Eds., Cameroon and Neighboring Basins in the Gulf of Guinea (Petroleum Geology tectonics Geophysics Paleontology and Hydrogeology), African Geoscience, 2008, pp. 13-37.

14. P. M. Thibaut and P. Le Berre, "Recherche d'Argiles Pour Briques Dans la Région de Yaoundé, Douala et Edéa," Rapport 85CM065, 1985.

15. D. Njopwouo, "Minéralogie et Physico-Chimie des Argiles de Bomkoul et de Balengou (Cameroun). Utilisation Dans la Polymérisation du Styrène et Dans le Renforcement du Caoutchouc Naturel," Thèse d'Etat, Faculté des Sciences, Université de Yaoundé, 1984.

16. D. Njopwouo and R. Wandji, "Minéralogie de l'Argile Kaolinique de Bomkoul (Cameroun)," Revue de Sciences et Technique, Série des Sciences de la Terre, I 3-4, 1985, pp. 71-81.

17. D. Njopwouo and S. Kong, "Minéralogie de la Fraction Fine des Matériaux Argileux de Bomkoul et de Balengou (Cameroun)," Annales de la Faculté des Sciences, Série des Sciences Chimiques, I 1-2, 1986, pp. 17-31.

18. A. Elimbi and D. Njopwouo, "Firing Characteristics of Ceramics from the Bomkoul Kaolinite Clay Deposit (Cameroon)," Tile and Brick International, Vol. 18, No. 6, 2002, pp. 364-369.

19. A. M. M. Mpondo, "Cartographie des Affleurements de la Localité de Bomkoul (Sous-Bassin de Douala)," Université de Douala, Mémoire, 2010.

20. M. B. Mbog, "Etude Morphologique, Physico-Chimique et Minéralogique des Argiles de Bomkoul Dans le Sous-Bassin Sédimentaire de Douala-Cameroun," Université de Douala, Douala, 2010.

21. G. F. Ngon Ngon, J. Etame, M. J. Ntamak-Nida, M. B. Mbog, A. M. Maliengoue Mpondo, M. Gérard, R. Yongue-Fouateu and P. Bilong, "Geological Study of Sedimentary Clayey Materials of the Bomkoul Area in the Douala Region (Douala Sub-Basin, Cameroon) for the Ceramic Industry," Comptes Rendus Geoscience, Vol. 344, No. 6, 2012, pp. 366-376.doi:10.1016/j.crte.2012.05.004

22. W. Samba, "Etude Morphologique, Géotechnique et Minéralogique des Argiles de Missole 2 Dans le SousBassin de Douala-Cameroun," Université de Douala, Douala, 2010.

23. E. O. Logmo, "Etude Géologique, Minéralogique et Géochimique des Argiles de Missole II Dans le Sous-Bassin de Douala (Cameroun)," Université de Douala, Douala, 2012.

24. G. F. Ngon Ngon, E. Bayiga, M. J. Ntamak-Nida, J. Etame, S. Noa Tang and R. Yongeu-Fouateu, "Trace Elements Geochemistry of Clay Deposits of Missole II from the Douala Sub-Basin in Cameroon (Central Africa): A Provenance Study," Sciences, Technologie et Développement, Vol. 13, No. 1, 2012, pp. 20-35.

25. G. F. Ngon Ngon, G. L. Lecomte Nana, R. Yongue Fouateu, G. Lecomte and P. Bilong, "Physicochemical and Mechanical Characterisation of Ceramic Materials Obtained from a Mixture of Silica, Feldspars and Clay Material of the Douala Region in Cameroon (Central Africa)," Advances in Ceramic Science and Engineering (ACSE), Vol., 2, No. 1, 2013, pp. 23-31.

26. R. Letouzey, "Atlas du Cameroun, Phytogéographie Camerounaise," Imprimerie Nationale Yaoundé, 1968.

27. A. K. Chakravorty and D. K. Ghosh, "Kaolinite-Mullite Reaction Series: The Development and Significance of a Binary Aluminosilicate Phase," Journal of the American Ceramic Society, Vol. 74, No. 6, 1991, pp. 1401-1406. doi:10.1111/j.1151-2916.1991.tb04119.x

28. P. Autret, "Latérites et Graveleux Latéritiques," Laboratoire Central des Ponts et Chaussées, 1983.

29. M. Younoussa, "Etude Géotechnique, Chimique et Miné- ralogique de Matières Premières Argileuses et Latéritiques du Burkina Faso Améliorées aux Liants Hydrauliques: Application au Génie Civil (Bâtiment et Route)," Thèse, Université de Ouagadougou, 2008.

30. NF P94-057, "Analyse Granulométrique des Sols. Méthode par Sédimentation," AFNOR, 1992.

31. NF P94-051, "Détermination des Limites d'Atterberg," AFNOR, 1993.

32. C. A. Jouenne, "Traité de Céramiques et Matériaux Mineraux," Septima, Paris, 1990.

33. S. Lee, Y. J. Kim and H. S. Moon, "Phase Transformation Sequence from Kaolinite to Mullite Investigated by an Energy-Filtering Transmission Electron Microscope," Journal of the American Ceramic Society, Vol. 82, No. 10, 1999, pp. 2841-2848.doi:10.1111/j.1151-2916.1999.tb02165.x

34. M. W. Carty, "The Colloidal Nature of Kaolinite," The Bulletin of the American Ceramic Society, Vol. 78, 1999, pp. 72-76.

35. E. A. Fitzpatrick, "Soils-Their Formation, Classification and Distribution," Longman, Berlin, 1983.

36. L. Cere and F. Mazel, "Caractérisation d'Argiles," ENSCI Limoges, 1993.

37.   Y. Beron and P. Le Berre, "Guide de Prospection des Matériaux," Manuels et Methodes du BRGM, No. 5, 1983.

38.   V. Rigassi, "Bloc de Terre Comprimée," Manuel de Prospection, Vol. 1, Craterre EAG, 1995.

39.   N. Française, "Sols: Reconnaissance et Essais. Description-Identification-Dénomination des Sols," XP P94-011, 1999.

40.   A. A. Al-Rawas and M. Qamarouddin, "Construction Problems of Engineering Structures Founded on Expansive Soils and Rocks in Northern Oman," Building and Environment, Vol. 33, No. 2-3, 1998, pp. 159-171.

41.   A. Djedid, A. Bekkouche and A. M. Aissa Mamoune, "Identification and Prediction of the Swelling Behavior of Some Soils from the Tlemcen Region of Algeria," Bulletin des Laboratoires des Ponts et Chaussées, Vol. 233, 2001, pp. 69-77.

42.   O. Castelein, "Influence de la Vitesse du Traitement Thermique sur le Comportement d'un Kaolin: Application au Frottage Rapide," Thèse, Université de Limoges, 2000.

Chapter 4

EVOLUTION OF LATERITIC SOILS GEOTECHNICAL PARAMETERS DURING A MULTI-CYCLIC OPM COMPACTION AND CORRELATION WITH ROAD TRAFFIC

Meissa Fall[1], Déthiè Sarr[1], Makhaly Ba[1], Etienne Berbinau[2], Jean-Louis Borel[2], Mapathé Ndiaye[1], Cheikh H. Kane[1]

[1]Laboratoire de Mécanique et Modélisation, UFR Sciences de l'Ingénieur, University of Thies, Thies, Senegal

[2]RAZEL sa, Christ de Sarclay, 3 rue René razel, Orsay cedex, Orsay, France

ABSTRACT

Gravel lateritic soils are intensively used in road geotechnical engineering. This material is largely representative of engineering soil all around the tropical African Countries [1,2]. Gravel lateritic soils from parts of Burkina Faso and Senegal (West Africa) are used to determine the evolution of the geotechnical parameters from one to ten cycles of modified Proctor compaction. This test procedure is non-common for geotechnical purposes and it was found suitable and finally adopted to describe how these problematic soils behave when submitted to a multi-cyclic set of Modified Proctor compactions (OPM) [3,4]. On another hand, we propose a correlation between the traffic and the cycles of compaction considered as the repeated load. From that, this work shows the generation of active fine particles, the decrease of the CBR index and also the mechanical characteristics (mainly the Young Modulus, E) that contribute at least to the main deformation of the road structure.

INTRODUCTION

This paper is primarily intended to demonstrate that under unpredicted traffic and repeated loading, properties of gravel lateritic soils used as pavement layer can significantly change. According to [5-10], gravel lateritic soils are very sensitive to an exceptional variation of stresses under which they are subjected in a pavement structural fill. Thus, it is expected that most of the physical and mechanical properties of gravel lateritic soils evolves during the design life.

It is then important to find an adequate method of testing that can deal with such behavior already known in the literature. It is then necessary to perform usual characterization tests on these kinds of materials by studying the evolution of their main properties under traffic such as gradation, plasticity, CBR (Californian Bearing ration), Los Angeles loss, Shear strength (UCT), etc.

To do this, tests are conducted so that they can simulate multi-cyclic axial loading generated by traffic loads. The first cycle of OPM compaction (cycle 1) corresponds to the specifications that are led to the initial design of pavement:

- Compaction at the Optimum Modified Proctor (OPM) and determination of the initial CBR value of the material that will have to support traffic.
- Determination during the same initial state of all physical and mechanical characteristics of materials, as reference values such as gradation, Atterberg limits, CBR, Los Angeles loss, Shear strength as Unconfined Compression Test characteristics (UCT), etc.
- And finally, perform multi-cyclic compaction procedure to determine soil characteristics at each cycle of compaction.

TEST PROCEDURE AND MATERIAL PROPERTIES

After complete characterization of a gravel lateritic specimen from Burkina Faso (between Boromo and Bobo Dioulasso mainly used for the design of this West African International Road) and Senegal (in the western part of the country, as Yenne and Thiès), (sieve and hydrometer analysis, Atterberg limits, methylene blue, etc.), soils are compacted and subjected to mechanical tests at the Optimum Modified Proctor (OPM). Theses mechanical tests are essentially CBR tests, unconfined compression test and resistance to degradation by abrasion and impact in the Los Angeles machine. After the first cycle, the remaining material is used to perform exactly the same tests during the subsequent cycles (2nd, 3rd, ..., 10th cycles) (**Table 1**). The purpose of these tests is to compare the evolution of main properties (particle size distribution, CBR, Young modulus, etc.) with repeated cycles of compaction. Tables 2(a) and (b) below summarize the overall results:

Table 1. Values of material properties at cycle 0 (raw material), (The main Lateritic Soils used in this paper are sampled from Burkina Faso between Boromo and Bobo Dioulasso).

	(fines) % < 80 m	PI (%)
Pk 247	14	15
Pk 272 + 600	16	17
Pk 284	15	10
Pk 288	18	15
Pk 342	20	12

Table 2a. Summary of the test results depending on the soil provenance and the cycles of compaction.

		Soil Tests												
Soil provenance (Pk)	Cycles of Compaction	Compaction Modified Proctor		Grain size distribution			Atterberg Limits			Other Soil Characteristics					
		γd max. (kN/m³)	W opt. (%)	(f) % < 5 mm	(m) % < 2 mm	(f) % < 0.08 mm	LL (%)	LP (%)	PI (%)	γs (kN/m³)	VBS	CBR (%)	Evolution (%)	E (MPa)	Rc (UCT) (MPa)
	1	21.6	8.7	40.4	30.4	17.3	33.5	13.4	20.5	2.75	2.15	88	100	26.90	0.15
	2	22.8	7.7	43.7	14.4	23.1	35.9	14.2	21.7			93	106	16.96	0.12
	3	23.3	8.2	60.5	31.4	25.0	39.4	15.4	24.0			95	108	59.50	0.35
	4	22.6	10.1	55.4	31.5	27.2	41.4	16.6	24.8		1.7	101	115	31.50	0.16
Pk 247	5	23.2	10.3	60.0	24.8	29.9	43.3	17.4	26.0		1.65	99	113	44.90	0.35
	6	23.2	8.4	58.0	20.0	33.3	46.1	19.4	26.7			87	99	61.10	0.34
	7	22.2	9.4	69.5	24.4	37.1	48.0	20.3	27.6			54	61	57.56	0.31
	8	21.5	9.8	45.9	22.1	44.1	52.6	21.8	30.8			49	56	45.98	0.28
	9	21.2	9.3	42.9	17.2	45.2	56.0	24.7	31.3			30	34	34.25	0.15
	10	22.3	7.6	31.9	16.3	46.2	57.1	26.6	30.6			29	33	29.06	1.10
	1	23.5	8.1	23.7	10.0	20.3	43.5	14.2	29.3	2.81	2.15	96	100	19.45	0.23
	2	23.4	9.3	33.8	12.1	23.4	49.5	15.4	34.1	2.47	1.95	103	107	23.43	0.25
	3	23.5	8.6	72.7	34.4	25.5	46.8	16.6	30.2	2.45	1.8	110	115	26.76	0.36
	4	23.9	8.7	51.0	20.0	25.0	49.6	19.1	30.5		1.75	102	106	45.87	0.39
Pk 272+600	5	23.8	8.0	57.7	19.0	29.7	52.3	18.1	34.2		1.7	113	118	76.98	0.45
	6	24.4	7.8	57.3	21.8	31.5	52.7	18.1	34.6			76	79	40.50	0.37
	7	23.1	8.4	59.8	18.8	39.6	53.3	18.9	34.4			48	50	15.90	0.22
	8	23.4	8.7	61.0	26.5	40.0	57.0	22.5	34.5			47	49	16.98	0.18
	9	22.5	9.8	44.7	22.6	45.2	62.3	21.3	41.0			35	36	14.00	0.16
	10	22.3	7.6	34.4	19.6	47.0	64.0	24.0	40.0			36	38	14.50	0.16
	1	22,5	8.7	63.8	40.5	28.5	33.9	13.4	20.5	2.87	1.55	75	100	35.71	0.15
	2	22.7	7.7	71.4	54.7	40.0	41.8	16.0	25.8	2.32	1.25	84	112	54.17	0.28
	3	23.3	7.1	58.3	38.6	45.1	52.4	22.2	30.2	2.12	1.1	54	72		
	4	23.8	6.7			47.5						50.7	68		
Pk 284	5	23.7	6.7									56	75		
	6											54	72		
	7											48	64		
	8														
	9														
	10														

Lateritic Soils sampled Between Boromo and Bobo Dioulasso (Burkina Faso)

Table 2b. Summary of the test results depending on the soil provenance and the cycles of compaction. * (Empty cells indicate insufficient quantity of materials for further testing. Multi-cyclic compaction uses a large amount of material per cycle. In this case, several samples were compacted at the same water content in order to provide enough amount of material for each cycle).

Soil provenance (Pk)		Cycles of Compaction	Soil Tests: Compaction Modified Proctor		Grain size distribution			Atterberg Limits			Other Soil Characteristics					
			$\gamma_{d\ max}$ (kN/m³)	W_{opt} (%)	(f) % < 5 mm	(m) % < 2 mm	(f) % < 0.08 mm	LL (%)	LP (%)	PI (%)	γ_s (kN/m³)	VBS	CBR (%)	Evolution (%)	E (MPa)	R_c (UCT) (MPa)
Pk 288	*Lateritic Soils sampled Between Boromo and Bobo Dioulasso (Burkina Faso)*	1	23.0	7.1	58.4	15.9	26.2	34.7	11.8	22.9	2.73	2.25	81	100	15.60	0.18
		2	22.6	9.3	43.4	19.7	30.3	42.8	16.4	26.4	2.5	1.9			23.15	0.20
		3	23.6	8.1	44.5	22.1	31.0	43.8	17.4	26.5	2.39	1.8			15.60	0.18
		4	24.2	7.6	44.0	21.0		47.4	18.1	29.3		1.75			32.30	0.38
		5	24.0	8.6	60.1	25.4		49.6	19.4	30.2					32.40	0.32
		6	22.5	11.8	57.0	19.8									28.70	0.32
		7	22.2	9.4	58.9	18.1									25.32	0.35
		8	21.5	9.8	44.7	20.4									20.50	0.25
		9	21.2	9.3	39.7	14.6									22.30	0.11
		10	20.4	9.3	33.6	18.6									18.15	0.12
Pk 342		1	23.4	7.3	43.7	33.7	28.2	31.3	15.5	15.8	2.87	1	84	100	33.33	0.11
		2	23.4	7.4	50.7	39.7	35.3	47.4	19.1	28.3	2.71	1.25	90	107	25.00	0.11
		3	23.3	5.5	50.4	38.8		61.8	21.1	40.6	2.33	0.9	97	115		
		4	23.7	5.7									105	125		
		5	23.6	6.8									95	113		
		6														
		7														
		8														
		9														
		10														

INTERPRETATION OF RESULTS

Generation of Fine particles and Changing in Characteristics of Consistency

As shown by figures below, the transition between first to 10th cycles contributes to a strong generation of fine particles, as well as a gradual increase of plasticity (**Figure 1**). The amount of fines particles (% < 80 μm) increases from 17% (which is the limit generally accepted for such materials) for the first cycle and reaches 46% for the 10th cycle. From the first to the 10th cycle, plasticity of materials also changes from 21% to 31% for the sample of Pk 247 and from 29% to 40% for the sample of Pk 272 + 600.

The **Figure 2** gives the results of Los Angeles tests performed on gravel lateritic soils samples. The test was conducted in a particular procedure that is "unconventional". In the case of the strict application of the standard, the test is performed in the fraction 10/14 with a mass of test sample of 5 kg. For our purposes, we took care to fill the hollow steel cylinder with the total fraction of the material without any selection. This procedure allows testing the total mass of the initial material without any selection and therefore allows completing **Figure 3** showing the generation of fine particles and changes in plasticity. Since the test measures the resistance to degradation by abrasion and impact of the material in a rotating steel drum containing a specified number of steel balls, results show a strong increase of percent loss by abrasion and impact as the number of cycle increases. In this sense, both coarse and fine aggregates fragment extensively during the test. This further demonstrates the problematic behavior of all gravel lateritic soils related in the literature [10].

Comparison with the Specifications in the Western African Area (West African Standards—WAS)

From **Figure 3** we can remark that, at the end of compaction cycles, materials tested are outside of specifications for the plasticity index and the amount of fine particles (<80 mm) as required by specifications.

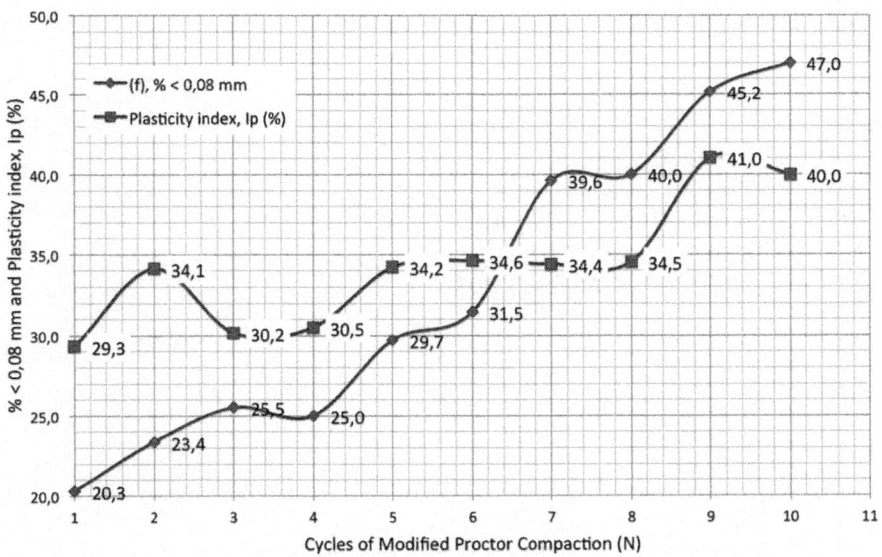

Figure 1. Evolution des fines (% < 80 m) et de la plasti-cité (PI) (Pk 272 + 600).

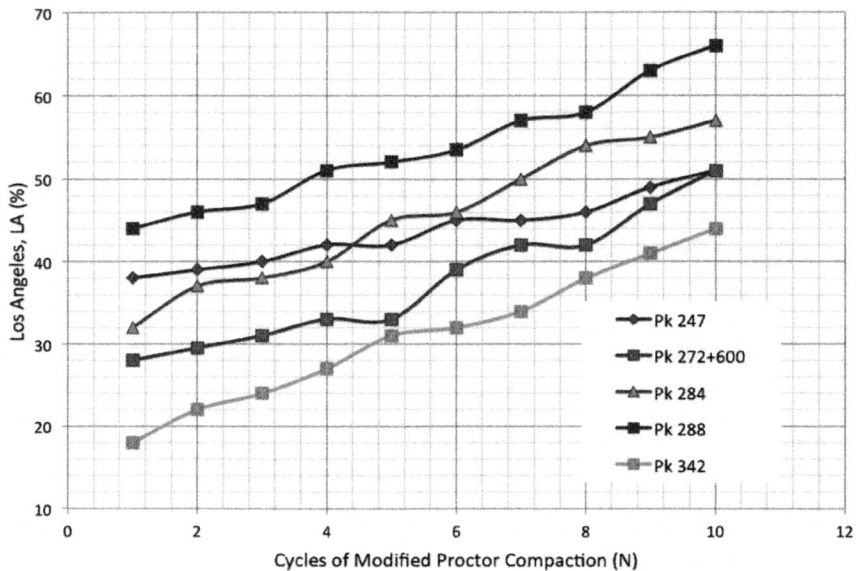

Figure 2. Evolution of percentage loss by abrasion.

PK 247

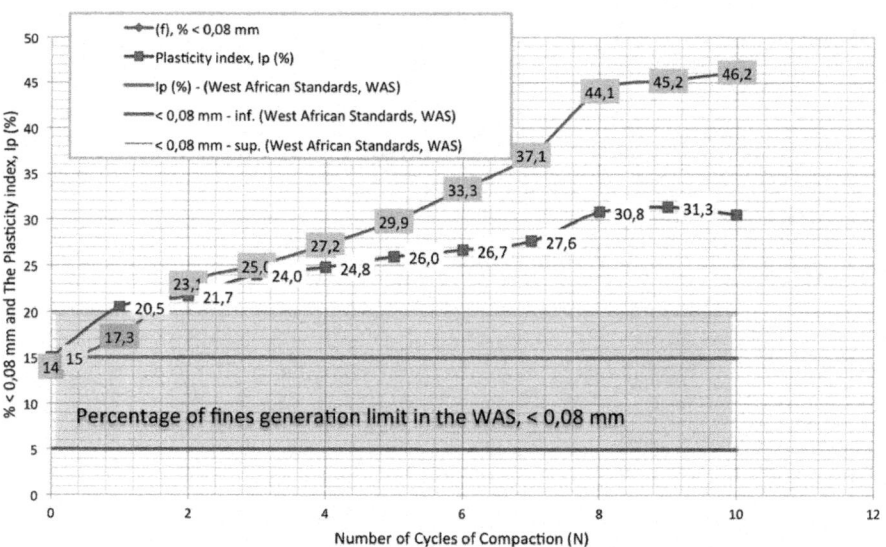

Figure 3. Comparison between results (<0.08 mm et PI (%)) and specification of the WAS.

Figure 4 shows the variation of CBR values with the cycles of modified Proctor compaction. **Table 3**below reminds technical recommendations contained in current textbooks approved by the CEBTP, the BCEOM and the LCPC [11] for the use of gravel lateritic soils as base courses and in the case of a T_1 to T_2 traffic level:

Evolution of the CBR Values

CBR is analyzed in several ways (**Figure 7**):

- In gross value, the CBR is changing slightly for all materials up to the 5th cycle. This trend towards material stiffening is well known. Fall et al. [10] underlines that behavior and attributes it to the fact that the soil is becoming denser during the first cy cles. It gradually changes from a loose state to a dense state. Air void between coarse grains tends to be reduced and filled by fine particles generated by the breaks of the material.

- The trend to the fifth cycle is to increase the CBR, which passes from a reference value of 100% and goes up to 118% or 113%. In gross value, the CBR increases from 88% to 101% and from 96 to 113%.

- After the fifth cycle, the CBR begins to drop strongly and eventually reaches extremely low values such as 29% and 36% (sometimes approaching 67%) for the gravel lateritic base course.

Trends explained in figures 5 and 6 are much clearer in **Figure 7** where the material stiffening is more perceptible. The stiffness increases from 0 to 5 cycles and then decreases considerably after the fifth cycle.

Note

Whatever the type of correlations made on the basis of CBR, we should have, in all cases, moduli that drop significantly when the number of cycles increases. In these cases, the design of pavement base courses should lead to a significant increase in thicknesses.

Table 3. Specification for a base layer for traffic $T_1 - T_2$.

	$CBR_{4d\ imbibition}$ at 95% OPM	PI (%)	% inf. at 80 μm (%)
CEBTP	80	<15	4 à 20
CEBTP-LCPC	80	<15	<15
CEBTP-BCEOM	80	<15	<15

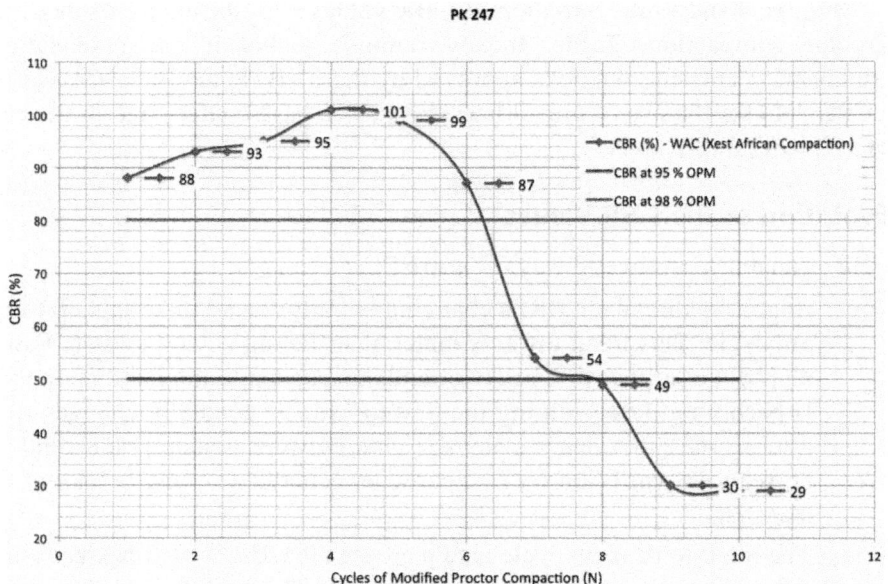

Figure 4. Comparison of CBR values with the requirements of the WAS.

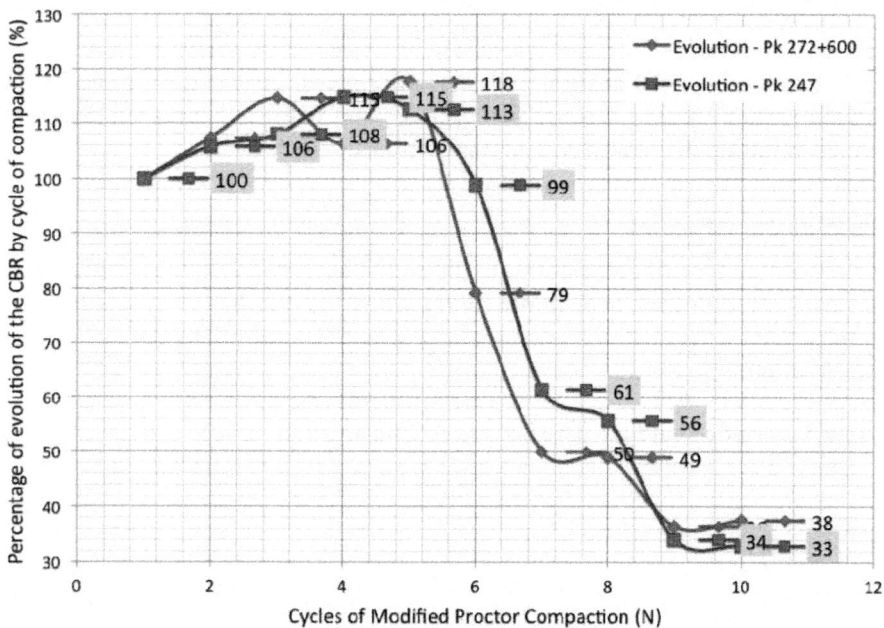

Figure 5. Percentage of evolution of the CBR values with cycles of compaction.

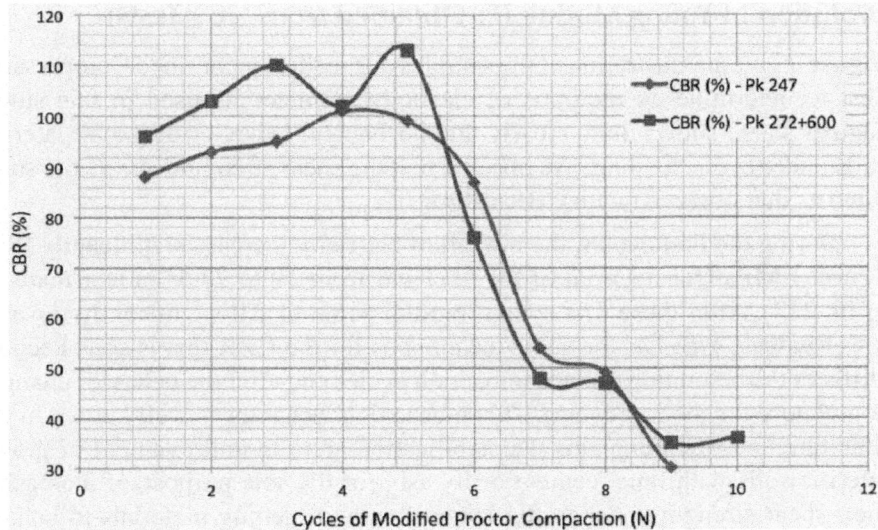

Figure 6. Evolution CBR value with cycle of compaction.

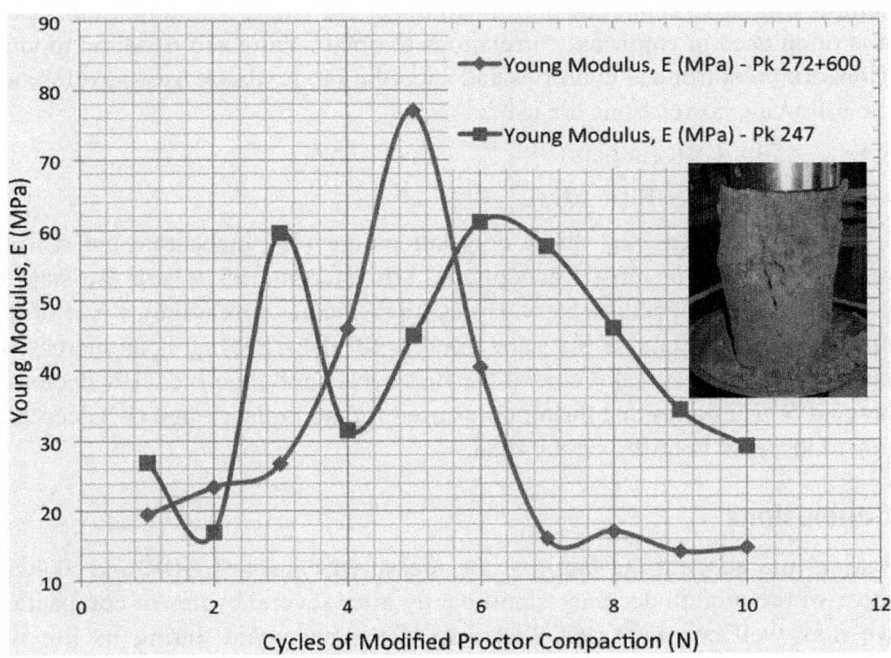

Figure 7. Evolution of the young modulus.

Evolution of Young Moduli (E: Obtained from UCT Tests)

Figure 7 give an illustration of the samples during the Unconfined Compression Test to determine de modulus of elasticity of materials used in this study. **Figure 8** also shows quite clearly the mechanical behavior of gravel lateritic soils under cyclic loading. We observe that the shear strength follows the same trend as that observed with CBR values.

During the first cycles, the moduli of elasticity increase significantly from 17 to 61 MPa (for material of Pk 247) and from 19 to 77 MPa (for material of Pk 272). After these first cycles, moduli begin to fall significantly towards lower values. This has the same meaning as for the CBR that is soils become stiffer at the beginning of the compaction cycles and after the behavior changes completely for the last cycles. This mechanical behavior is well known in the literature and often explains the stabilization and the improvement of gravel lateritic soils with lime, cement or fly ash, for the sole purpose of increasing their shear strength under traffic loading without getting materials to behave as a slab.

CBR is an important parameter in pavement design if unconfined compression tests cannot be performed to get the Young's moduli. In this case, it is often used in empirical correlations to obtain static and dynamic moduli. Thus, for most tropical countries and according to textbook used as reference, the following correlations are used:

$\sqrt{E_{static}} = 50 \times CBR$ (in bars)

$\sqrt{E_{dynamic}} = 100 \times CBR$ (in bars).

Although often used, these correlations are very inaccurate but still are references today in most francophone African countries where the state of the research is still rudimentary. By using the same correlation as part of this project, we get of course the same trend as for the CBR that is an increase of modulus towards a peak value at the first cycles and then the CBR decreases beyond. This implies that the modulus used for the initial design (E_0) decreases due to increase in traffic on the road.

Conclusions

Taking into account the fact that the measured values of CBR and likewise those of the moduli decrease significantly after several cycles of compaction, we may well conclude that thickness of the pavement during its life will also differ significantly from the initial designed thickness. The immediate conclusion to this is that:

- The design life of the pavement is significantly reduced and lead to premature ruin of the structurel Initial thicknesses should be higher if the designer was well aware of these behaviors.

CORRELATION BETWEEN ENERGY OF COMPACTION AND ENERGY OF TRAFFIC

Energy of Compaction

The energy of compaction is given by:

$$E_C = \frac{N \times m \times g \times h}{V_m}$$

N: number of blows; m: mass of the hammer; g: acceleration due to gravity; h: height of drop of the hammer and Vm: volume of the Proctor or CBR mold.

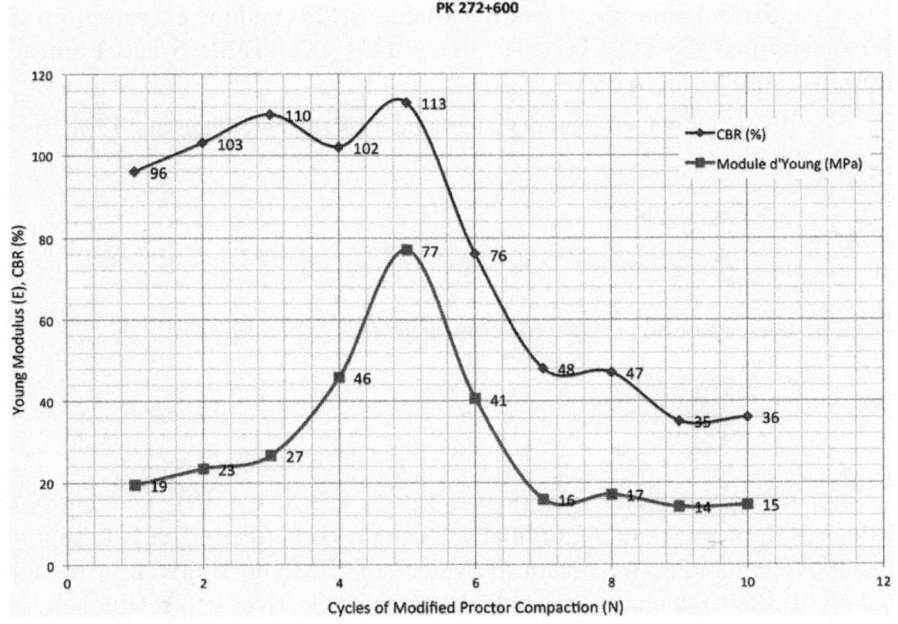

Figure 8. Evolution of design parameters (young's modulus and CBR).

Energy Due to Traffic

By analogy with the energy of compaction (**Table 4**), the energy of traffic can be expressed as below:

$$E_t = \frac{K \times g \times h}{V}$$

K is then defined by:

$$K = Q \times TJM$$

Q is the standard axle load converted to kgG is the acceleration due to gravity.

h (m) is the thickness of the pavement, $V = h \times \pi r^2$ is the volume of materials involved under the standard axle (in m^3).

After simplification E_t becomes:

$$E_t = \frac{Q \times TJM \times h}{V}$$

So in that relationship, the only variable is TJM and the energies of traffic are determined for TC_0, TC_1, TC_2, TC_3, TC_4, TC_5 (**Table 5** and **Figure 9"** **target="_self"> Figure 9**).

For given values of x and y, we can directly calculate the TJM by the equation below:

$$TJM = \frac{y \times log_{10} \, y \times S}{Q \times g}$$

The corresponding energy of compaction is expressed as:

$$E_C = \frac{y \times log_{10} \, x \times S'}{V_m}$$

For four given values of x ($x_1 = 2$, $x_2 = 5$, $x_3 = 8$, $x_4 = 10$), we identify the values of the corresponding ordinates for the two energies (E_t et E_c). Applying the above equations, we obtain the values of TJM and E_c given in the table below (**Table 6**). k_t and k_c are calculated; the objective is to relate them in a relationship in order to achieve the correlation.

Let:

1 kt(1) = 5.4, kt(2) = 12.6 et kt(3) = 19.2 progression factors of Etl kc(1) = 23, kc(2) = 50; kc(3) = 66 progression factors of the energy of compaction.

These ratios are calculated as below:

$$\frac{K_C(1)}{K_t(1)} = \frac{K_C(2)}{K_t(2)} = \frac{K_C(3)}{K_t(3)}$$

$K_c = 4k_t$, let 1 TJM$_0$ = 10 and TJMi (i = 1, 2, 3...n)

1 E$_0$ = 8 761,5

Table 4. Summary of the parameter of the curve E$_c$.

Cycles	1	2	3	4	5	6	7	8	9	10
N	275	550	825	1100	1375	1650	1925	2200	2475	2750
E$_n$ (KJ)	2635	5276	7904	10,539	13,174	15,809	18,443	21,078	23,713	26,348
p	1	2	3	4	5	6	7	8	9	10

Table 5. Summary of the parameters for drawing the curve E$_t$.

Classes TCi	TC$_0$	TC$_1$	TC$_2$	TC$_3$	TC$_4$	TC$_5$
TJM (*in heavy trucks*)	2	14	27	68	164	342
E$_t$ (kJ)	51,908	363,354	700,754	1,764,862	4,256,431	8,876,217
Progression factor	1	7	13.7	34	82	171

Table 6. Increase in the number of heavy load vehicles and energy of compaction.

TJM	10	54	126	192
k$_t$	1	5.4	12.6	19.2
Ec (kJ)	8761	203,435	438,073	639,587
k$_c$	1	23	50	73

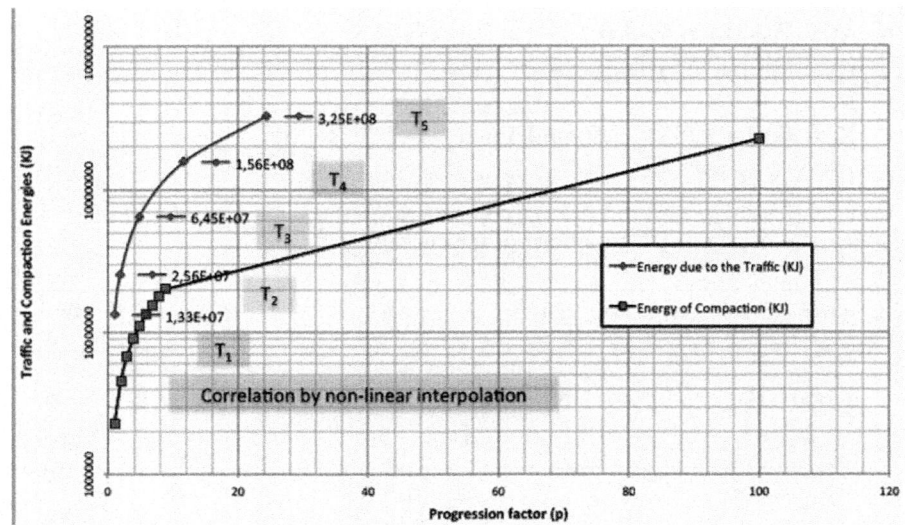

Figure 9. Curves of traffic energy and compaction energy vs progression factor.

We can write :

$$\frac{E_C(i)}{E_0} = K_C = 4Kt = 4.\frac{TJM_i}{TJM_0}$$

Hence,

$$E_C(i) = \frac{4E_0}{TJM_0}.TJM_i$$

In this formula, the energy is expressed in k_j. The formula reflects a geometric series with a common ratio expressed as below:

$$q = \frac{4E_0}{10}$$

This result allows the designer to assume the desired traffic and then deduce the corresponding energy of compaction.

Conclusions

We note well that the multi-cyclic compaction simulate exactly the effect of traffic loading. In this sense, observe that increase in traffic can be simulated by an increase in the compacting cycle. At the end of compaction, the traffic reaches very high level (T_4 to T_5).

CONCLUSIONS

Results show clearly that under multi-cyclic compaction, gravel lateritic soils generate fine particles, which increase their plasticity and drop their CBR value. Similarly, it is clearly shown that multi-cyclic compaction simulates well the effect of traffic by allowing reaching its expected level, which is highest traffic level at the end of the cyclic compaction.

ACKNOWLEDGEMENTS

The authors would like to acknowledge again the Entreprise Jean Lefebvre sa for their guidance and valuable input in this research project. The lateritic soils from Burkina Faso (from Boromo to Bobo Dioulasso) were collected by the help of the Entreprise Razel and sended to us by the Razel's Team of Bamoko (Mali). By this, we thank so much all of them for their ability of making the in situ samplings as well as we desire for the laboratory's tests. The large amount of materials used for this work has made the L2M laboratory's technicians as busy as they never be, the authors would like to thank them for their important contributions.

REFERENCES

1. A. Remillon, "Les recherches Routières Entreprises en Afrique D'expression Française, Application à la Conception et au Renforcement des Chaussées économiques, " ITBTP, 1967.

2. M. D. Gidigasu, "Laterite Soil Engineering-Pedogenesis and Engineering Principles," Elsevier Scientific Publishing Company, Amsterdam, 1976.

3. T. Aashto, "Standard Method of Test for MoistureDensity Relations of Soils Using a 2.5-kg (5.5-lb) Rammer and a 305-mm (12-in.) Drop. In Standard Specifications for Transportation Materials and Methods of Sampling and Testing," American Association of State Highway and Transportation Officials (AASHTO), Washington DC, 1999.

4. T. Aashto, "Standard Method of Test for the California Bearing Ratio," In Standard Specifications for Transportation Materials and Methods of Sampling and Testing, American Association of State Highway and Transportation Officials (AASHTO), Washington DC, 1993.

5. H. R. Sreekantiah, "Laterites and Lateritic Soils of West Coast in India," Southeast Asian Geotechnical Conference, Bangkok, Vol. 1, 27 December 1987, pp. 159-170.

6. A. Sridhan, "General Report: Engineering Properties of Tropical Soils," International Conference on Geomechn in Tropical Soils, Singapore, 12 December 1988.

7.   G. Sweere, P. Galjaard and T. Tjong, "Engineering Behaviour of Laterites in Road Constructions," International Conference on Geomechn in Tropical Soils, Singapore, Vol. 1, 1988, pp. 421-427.

8.   B. A. Vallerga, et al., "Engineering Study of Laterites and Lateritic Soils in Connection with Construction of Road, Highways and Airfields," Country Report, AID/CSD- 1810, 1969, p. 165 and appendices.

9.   L. W. Ackroyd, "Engineering Classification of Some Western Nigerian Soils and Their Qualities in Road Building," Road Research Laboratory, British, 1959.

10. M. Fall, "Caractérisation et Identification mé- canique de Trois Graveleux Latéritiques du Sénégal Occidental: Application au domaine routier," Thèse de Doctorat de l'INPL en Génie Civil et Minier de l'INPL-Nancy France, 1993, p. 277.

11. Secrétariat d'Etat aux Affaires Etrangères – CEBTPBCEOM (1984) - Guide pratique de dimensionnement des chaussées dans les pays tropicaux, CEBTP ; 2 ème édition: tome 1: Politique et économie routière ; 2 ème édition, 1991; tome 2: Etudes techniques et construction ; 2 ème édition, 1991; tome 3: Entretien et gestion des routes ; 2 ème édition, 1991.

Chapter 6

GEOCHEMICAL SPECIATION AND RISK ASSESSMENT OF HEAVY METALS IN SOILS AND SEDIMENTS

Santosh Kumar Sarkar[1], Paulo J.C. Favas[2, 3], Dibyendu Rakshit[1], and K.K. Satpathy[4]

[1]Department of Marine Science, University of Calcutta, Calcutta, West Bengal, India

[2]Department of Geology, School of Life Sciences and the Environment, University of Trás-os-Montes e Alto Douro, Vila Real, Portugal

[3]IMAR-CMA Marine and Environmental Research Centre, Faculty of Sciences and Technology, University of Coimbra, Coimbra, Portugal

[4]Indira Gandhi Centre for Atomic Research, Environment and Safety Division, Kalpakkam, Tamil Nadu, India

INTRODUCTION

Heavy metal pollution is a serious and widely environmental problem due to the persistent and non-biodegradable properties of these contaminants. Sediments serve as the ultimate sink of heavy metals in the marine environment and they play an important role in the transport and storage of potentially hazardous metals. They are introduced into the aquatic system as a result of weathering of soil and rocks, from volcanic eruptions and from a variety of human activities involving mining, dredging, processing and use of metals and/or substances containing metal contaminants. Heavy metals entering natural water become part of the water-sediment system and their distribution processes are controlled by a dynamic set of physicochemical interactions and equilibria. The properties of metals in soils and sediments depend on the physiochemical form in which they occur [1]. Heavy metals are distributed throughout soil and sediment components and associated with them in various ways, including adsorption, ion exchange, precipitation and complexation and so on [2]. Changes in environmental conditions, such as temperature, pH, redox potential and organic ligand concentrations, can cause metals to be released from solid to liquid phase and sometimes cause contamination of surrounding waters in aquatic systems [3]. They are not permanently fixed by soil or sediment. Therefore, it cannot provide sufficient information about mobility, bioavailability and toxicity of metals if their total contents are studied alone.

Natural and anthropogenic activities have the capacity to cause changes in environment conditions, such as acidification, redox potential, or organic ligand concentrations, which can remobilize contaminated soils and sediments releasing the elements from soils and sediments and pore water to the water column resulting contamination of surrounding waters. Daily tidal currents, wind energies, and storms in coastal and estuarine systems can cause periodical remobilization of surface sediments [4]. More turbulent conditions, such as seasonal flooding or storms, or bioturbation, due to feeding and movement of benthic organisms, can expose anoxic sediments to oxidant conditions. In addition, activities such as dredging result in major sediment disturbances, leading to changes in chemical properties of sediment [5].

The remediation of heavy metal pollution is often problematic due to their persistence and non-degradability in the environment. As a sink and source, soils and sediments constitute a reservoir of bioavailable heavy metals and play a significant role in the remobilization of contaminants in the aquatic systems under favorable conditions. Such potential of sediment for being a sink as well as a source of contaminant can make sediment chemistry and toxicity key components of the quality of aquatic system. Much concern has been focused on the investigation of the total element contents in soils and sediments. However, it cannot provide sufficient information about mobility, bioavailability and toxicity of elements and thus may not be able to provide information about the exact dimension of pollution. The data on total contents of metals are quite insufficient to estimate the possible risk of remobilization of total metals under changing environmental conditions and potential uptake of liberated metals by biota and thus the determination of different fractions assume great importance. This has been described as "speciation" [6]. Since each form have different bioavailability and toxicity, the environmentalists are rightly concerned about the exact forms of metal present in the aquatic environment.

The concept of speciation dates back to 1954 when Goldberg introduced the concept of speciation to improve the understanding of the biogeochemical cycling of trace elements in seawater. Kinetic and thermodynamic information together with the analytical data made it possible to differentiate between oxidized versus reduced, complexed or chelated versus free metal ions in solution and dissolved between particulate species. Florence [7] has defined the term speciation analysis as the determination of the individual physicochemical forms of the element, which together make up its total concentration in a sample. According to Lung [8], speciation analysis involves the use of analytical methods that can provide information about the physicochemical forms of the elements. Schroeder [9] distinguishes physical speciation, which involves differentiation of the physical size or the physical properties of the metal, and chemical speciation, which entails differentiation among the various chemical forms. The main objective of measuring metal species relates to their relative toxicities to aquatic biota. The second and long term aim of speciation studies is to advance an understanding of metal interactions between water and bed sediments in an aquatic ecosystem. In the last decade researchers have followed different sequential extraction techniques for the fractionation of metals in sediments of different river systems. Rauret et al. [10] studied the speciation of copper and lead in the sediments of River Tenes (Spain) while Pardo et al. [11] studies the speciation of zinc, cadmium, lead, copper, nickel and cobalt in the sediments of Pisuerga River, Spain, in order to establish the extent to which these are polluted and their capacity to remobilization. Jardo and Nickless [12] investigated the chemical association of zinc, cadmium, lead and copper in soils and sediments of England and Wales. In most samples, these four metals were associated with all the chemical fractions. Tessier et al. [13] studied speciation of cadmium, cobalt, copper, nickel, lead, zinc, iron and manganese in water and sediments of St. Fransois River, Quebec, Canada. Elsokkary and Muller [14] studied speciation of chromium, nickel, lead and cadmium in the sediments of Nile River, Egypt, reporting that a high proportion of chromium, nickel and lead is bound to organic material and sulphides, while cadmium is bound to carbonate fraction. Ure [15] and Rauret [16] have reviewed the chemical extraction procedures used for heavy metal determinations in contaminated soils and sediments. Owing to the need for validation of extraction schemes, the EC Measurement and Testing Programme (formerly BCR) has organized a project for improving the quality of determinations of extractable heavy metals, where development and validation of extraction procedures has been discussed [17, 18].

The present article aims to summarize the potentials of sequential extraction technique adopting different analytical protocols for gaining information on the mobility and dynamics of operationally determined chemical forms of heavy metals in soils and sediments. The BCR (Community Bureau of Reference, now superseded by the Standards, Measurement and Testing Programme of the European Community) procedure has been illustrated considering the case study of Ganges (Hugli) River Estuary and adjacent Indian Sundarban mangrove wetland (a UNESCO World Heritage Site), northeastern part of the Bay of Bengal. In addition, the authors also evaluate the modified BCR sequential extraction technique as devised by various scientists, the risk assessment code (RAC) as well as assessment of toxicity comparing with sediment quality guidelines. The RAC classification is based on the strength of the bond between the metals and the different geochemical fractions in sediments or soils and the ability of metals to be released and enter into the food chain.

SEQUENTIAL EXTRACTION: MERITS AND DEMERITS

The sequential extraction provides more or less detailed information concerning the origin, mode of occurrence, biological and physicochemical availabilities, mobilization and transport of heavy metals. The procedure stimulates the mobilization and retention of these species in the natural environment using changes in environmental condition such as pH, redox potential and degradation of organic matter [16]. A series of reagents is applied to the sample, increasing the strength of the extraction at each step, in order to dissolve the trace metal present in different sediment phases. The extractants are inert electrolytes, weak acids, reducing agent, oxidizing agents and strong mineral acids [19].

The 3-stage sequential extraction procedure proposed by the European Community Bureau of Reference (BCR) was developed in an attempt to standardize the various schemes described in the literature [2, 20,21], since the use of different procedures, varying in the number of steps, types of reagents and extraction condition. Hindered comparison of results obtained in the many studies of heavy metals chemical fractionation in environmental samples [22].

The BCR methods has been widely adapted by various authors, and applied to a range of type of solid sample including fresh water sediment [23-25], salt water sediment [26-28], sewage sludge and particulate matter [29-31]. This scheme enables us to associate the meals with one of the following four geochemical phases:

I. Acid-soluble phase: This phase is made up of exchangeable metals and others bound to carbonates that are able to pass easily into the water column, for example, when the pH drops. It is the fraction with the most labile bond to the soil/sediment and, therefore, the most dangerous for the environment.

II. Reducible phase: This phase consists of metals bound to iron and manganese oxides that can be released if the sediment changes from the oxic to the anoxic state, which could be caused, for example, by the activity of microorganisms present in the soils/sediments.

III. Oxidisable phase: This shows the amount of metal bound to organic matter and sulphides, which can be released under oxidizing conditions. Such conditions can occur, for example, if the sediment is resuspended (by dredging, currents, flooding, tides, etc.) and the sediment participles come into contact with oxygen-rich water.

IV.   Residual phase : Lithogenous and inert (Non-bioavailable).

The heavy metals in the soils and sediments are bound to different fractions with different strengths, the value can, therefore, give a clear indication of soil and sediment reactivity, which in turn assess the risk connected with the presence of heavy metals in a terrestrial or aquatic environment. The rationale of the sequential extraction procedure is that each successive reagent dissolves a different component, which can content heavy metals within their crystalline structures. Under natural conditions, metals in minerals are unlikely to experience significant release over the time frames of interest [32, 33].

ANALYTICAL PROTOCOLS FOR SEQUENTIAL EXTRACTION

In recent years a great number of papers have been published on various analytical techniques proposed for the fractionation analysis of trace elements in various environmental samples (soils, sediments, etc.). An approach that has been found to be preferable is the fractionation of heavy metal into operationally defined forms under the sequential action of different extractants [2]. Selective extractants, used in sequential extraction procedures, are aimed at the simulation of natural conditions whereby metals associated with certain soil (sediment) components can be released. For example, changes in the ionic composition affecting adsorption–desorption reactions or a decrease in pH may lead to the release of metals, retained on a matrix by weak electrostatic interactions or co-precipitated with carbonates ("exchangeable" and "acid soluble" forms). Decreasing the redox potential can result in dissolution of oxides, unstable under reducing conditions, and liberation of scavenged metals ("reducible" forms). Changes in oxidizing conditions may cause the degradation of organic matter and release of complexed metals ("oxidizable" forms). Finally, the destruction of primary and secondary mineral lattice releases heavy metal retained within the crystal structure, e.g., due to isomorphous substitution ("residual" forms) [2]. The nominal "forms" determined by operational fractionation can help to estimate the amounts of total metals in different reservoirs which could be mobilized under changes in the chemical properties of the soil [34]. Since the 1970s a considerable number of extraction procedures have been proposed for determining the forms of heavy metal [2, 35-39]. Most of these procedures are based on the scheme of Tessier et al. [2]. Although most of the extracting reagents were originally used in the chemical analysis of soils, the procedures proposed have been tested on a wide variety of contaminated environmental samples—sediments, road dust, sewage sludge, etc.

Sequential extraction can be useful to have an operational classification of metals in different geochemical fractions [2] which is the most reliable criteria to quantify the potential effect of soil/sediment contamination by heavy metals. This can provide information about the identification of the main binding sites, the strength of element binding to the particulates and the phase associations of trace elements in soil/sediment. Following this basic scheme, some modified procedures with different sequences of reagents or operational conditions have been developed [40-43]. Considering the diversity of procedures and the lack of uniformity in different protocols, a European Community Bureau of Reference (BCR, now the European Community Standards Measurement and Testing Program) method was proposed [6] and was applied by a large group of researchers [31, 44-47]. In this study, we followed the sequential extraction procedure proposed by the European Union's Standards, Measurements and Testing program [3].

MODIFIED BCR SEQUENTIAL EXTRACTION PROCESS

As discussed above it is evident that sequential extraction provides valuable information regarding identification of main binding site, the strength of the element binding to the particulates and the phase associations of heavy metals in sediments. However, various complicated sequential extraction procedures were experimented to provide more detailed information regarding different metal phase associations [2, 48, 49]. A wide range of techniques is available whereby various extraction reagents and experimental conditions are used. These techniques involve a 5-step [2], 4-step (BCR, Bureau Commune de Reference of the European Commission), 6-step [50] and 7-step [51, 52] extraction, and are thus becoming popular methods to be used for sequential extraction [53, 54]. Following this basic scheme, some modified procedures with different sequences of reagents or operational conditions have been developed [40-43].

Several sophisticated instruments have been used for the determination of heavy metals contents in marine environments. These include; flame AAS [55, 56], atomic fluorescence spectrometry [57], anodic stripping voltammetry [58, 59], ICP-AES [60] and ICP-MS [61, 62].

Heavy metal mobility and bioavailability depend strongly on their chemical and mineralogical forms in which they occur [63]. Several speciation studies have been conducted to determine study different forms of heavy metals rather their total metal content. These studies reveal the level of bioavailability of metals in harbour sediments and also confirm that sediments are indicators of heavy metal pollution in marine environment [64-67].

Since the early 1980s and 1990s sequential extraction methodology has been developed to determine speciation of metals in sediments [2, 68] due to the fact that the total concentration of metals often does not accurately represent their characteristics and toxicity. In order to overcome the above mentioned obstacles it is helpful to evaluate the individual fractions of the metals to fully understand their actual and potential environmental effects [2]. To date, strong acid digestion is used often for the determination of total heavy metals in the sediments. However, this method can be misleading when assessing environmental effects due to the potential for an overestimation of exposure risk. Moreover, in order to determine the mobility of heavy metals in sediments, various sequential extraction procedures have been developed [69-71].

Among a range of available techniques using various extraction reagents and experimental conditions to investigate the distribution of heavy metals in sediments and soils, the 5-step Tessier et al. [2] and the 6-step extraction method, Kersten and Fronstier [50] were mostly widely used. Following these two basic schemes, some modified procedures with different sequences of reagents or experimental conditions have been developed [40-43]. Considering the diversity of procedures and lack of uniformity in different protocols, a BCR, Bureau Commun de Recherche (now called the European Community (EC) Standards Measurement and Testing Programme) method was proposed [6]. It harmonized differential extraction schemes for sediment analysis. The method has been validated using a sediment certified reference material BCR-701 with certified and indicative extractable concentration of Cd, Cr, Cu, Ni, Pb and Zn [72]. This method was applied and accepted by a large group of specialists [31, 44, 45, 47,73, 74] despite some shortcoming in the sequential extraction steps [75, 76].

Wang et al. [77] used a modified Tessier sequential extraction method to investigate the distribution and speciation of Cd, Cu, Pb, Fe, and Mn in the shallow sediments of Jinzhou Bay, Northeast China. This site was heavily contaminated by nonferrous smelting activities. They found out that the concentrations of Cd, Cu and Pb in sediments was to be 100, 73, 13 and 7 times, respectively, higher than the National guidelines (GB 18668-2002). The sequential extraction tests revealed that 39%-61% of Cd was found in exchangeable fractions. This shows that Cd in the sediments posed a high risk to the local environment. Copper and Pb were found to be at moderate risk levels. According to the relationships between percentage of metal speciation and total metal concentration, it was concluded that the distributions of Cd, Cu and Pb in some geochemical fractions were dynamic in the process of pollutants migration and stability of metals in marine sediments from Jinzhor Bay decrease in the order Pb>Cu>Cd.

Yuan et al. [78] applied BCR-sequential extraction protocol to obtain metal distribution patterns in marine sediments from the East China Sea. The results showed that both the total contents and the most dangerous non-residual fractions of Cd and Pb were extremely high. More than 90% of the total concentration of V, Cr, Mo and Sn existed in the residual fraction while more than 60% of Fe, Co, Ni, Cu, and Zn were mainly present in the residual fraction. Manganese, Pb, and Cd were dominantly present in the non-residual fractions in the top sediments.

Jones and Turki [79] worked on distribution and speciation of heavy metals in surface sediments from the Tees estuary, North East England. Tessier et al. [2] metal speciation scheme modified by Ajay and van Loon [80] was used for the study. They observed out that the sediments were largely organic-rich clayey silts in which metal concentrations exceed background levels, and which attain peak values in the upper and middle reaches of the estuary. Chromium, Pb and Zn were associated with the reducible, residual, and oxidizable fractions. Cobalt and Ni were not highly enriched while Cu is associated with the oxidizable and residual fractions. Cadmium is associated with the exchangeable fractions.

Pempkowlak et al. [81] investigated the speciation of heavy metals in sediments and their bioaccumulation by mussels. They used a 4-step sequential extraction procedure adapted from Forstner and Watmann [82]. Their investigation which was characterized by varying metal bioavailability was aimed at revealing differences in the accumulation pattern of heavy metals in mussel inhabiting that inhabit in sediments. The bioavailabilities of metals were measured using the contents of metals adsorbed to sediments and associated with Fe and Mn hydroxides. The biovailable fraction of heavy metals contents in sediments collected from Spitsbergen represented a small proportion (0.37% adsorbed metals and 0.11%, are associated with metals hydroxides). It was also revealed that the percentages of metals adsorbed and bound to hydroxides of the sediments ranged from 1 to 46% and 1 to 13%, respectively.

Wepener and Vermeulen [66] investigated on the concentration and bioavailability of selected metals in sediments of Richards Bay harbor, South Africa. Sequential extraction of sediments was carried out according to Tessier et al. [2] method. The following metals were investigated: Al, Cr, Fe, Mn, and Zn. Their studies revealed that metals concentrations in sediments samples varied only slightly between seasons, but showed significant spatial variation, which was significantly correlated to sediment particle size composition. Highest metal concentration was recorded in sites with substrates dominated by fine mud. Manganese and Zn had more than 50% of this concentration in reducible fraction while more than 70% of the Cr was associated with the inert fractions and the concentration recorded at some sites were still above action levels when considering only the bioavailable fractions. They also concluded that the concentration of Zn recorded was not elevated their results were compared with the historic data.

Coung and Obbard [54] used a modified 3-step sequential extraction procedure to investigate metal speciation in coastal marine sediments from Singapore as described by the European Community Bureau of Reference (ECBR). Highest percentages of Cr, Ni, and Pb were found in residual fractions in both Kranji (78.9%, 54.7% and 55.9% respectively) and Pulang Tokong (82.8%, 77.3% and 62.2% respectively). This means that these metals were strongly bound to sediments. In sediments from Kranji, the mobility order of heavy metals studied were Cd>Ni>Zn>Cu>Pb>Cr while sediments from Pulan Tekong showed the same order for Cd, Ni, Pb and Cr, but had a reverse order for Cu and Zn (Cu>Zn). The sum of the 4-steps (acid soluble + reducible + oxidizable + residual) was in good agreement with the total metal content, which confirmed the accuracy of the microwave extraction procedure in conjunction with the GFASS analytical method.

Fedotov et al. [83] applied a modified technique for accelerated fractionation of heavy metals in contaminated soils and sediments using rotating coiled columns. Rotating coiled columns (RCC) is valuable for the continuous-flow sequential extraction and can be successfully applied to the dynamic leaching of heavy metals from soil and sediments. This is a fluoroplastic or steel coil wound around a rigid cylindrical drum, which revolves about its axis and, at the same time, revolves around the central axis of the device called planet centrifuge. The stationary (liquid, solid, or heterogeneous) phase is retained in the column because of the centrifugal force field, and the mobile liquid phase is continuously pumped through the column. A solid sample was retrieved in the rotating column as the stationary phase under the action of centrifugal forces while different elements (aqueous solution of complexing reagents, mineral salts and acids) were continuously pumped through. This procedure developed is time saving and requires only 4-5 hr instead of the several days needed for individual sequential extraction. Losses of solid sample are minimal. Further studies are needed to better estimate the reproducibility of the technique.

Nemati et al. [84] used a modified BCR sequential extraction procedure (SEP) in combination with ICP-MS to obtain the metal distribution patterns in different depths of sediments from Sungai Buloh, Selangor, Malaysia. The results showed that heavy metal contaminations at Sungai Buloh River sediments were more severe than at other sampling sites, especially for Zn, Cu,

Ni and Pb. Nevertheless, the element concentrations from top to bottom layers decreased predominantly.

Mossop et al. [85] compared of original and modified BCR sequential extraction procedures for the fractionation of Cu, Fe, Pb, Mn and Zn in soils and sediments. The procedures were applied to five soil and sediment substrates: a sewage sludge-amended soil, two different industrially contaminated soils, river sediment and intertidal sediment. Extractable Fe and Mn concentrations were measured to assess the effects of the procedural modifications on dissolution of the reducible matrix components. Statistical analyses (two-tailed t-tests at 95% confidence interval) indicated that recovery of Fe in step 2 was not markedly enhanced when the intermediate protocol was used. However, significantly greater amounts were isolated with the revised BCR scheme than with the original procedure. Copper behaved similarly to Fe. Lead recoveries were increased by use of both modified protocols, with the greatest effect occurring for the revised BCR extraction. In contrast, Mn and Zn extraction did not vary markedly between procedures. The work indicates that the revised BCR sequential extraction proves better attack on the Fe-based components of the reducible matrix for a wide range of soils and sediments.

SEQUENTIAL EXTRACTION OF METALS IN SEDIMENTS OF THE HUGLI RIVER ESTUARY AND INDIAN SUNDARBAN WETLAND: A CASE STUDY

Materials and Methods

Sample Collection and Sediment Quality Analysis

The delta region formed by Hugli (Ganges) River Estuary (HRE) and is famous for its luxuriant mangrove vegetation, known as Sundarban wetland, acclaimed as UNESCO World Heritage Site for its capacity of sustaining an excellent biodiversity. The wetland is characterized by a complex network of tidal creeks, which surrounds hundreds of tidal islands exposed to different elevations at high and low semi-diurnal tides. This is one of the most sensitive and vulnerable ecosystems in the world and suffers from environmental degradation due to rapid human settlement, tourism and port activities, operation of mechanized boats, deforestation, and increasing agricultural and aquaculture practices. The ongoing degradation is also related to huge siltation, flooding, storm runoff, atmospheric deposition, and other stresses resulting in changes in water quality, depletion of fishery resources, choking of river mouth and inlets, and overall loss of biodiversity. Moreover, the rapid economic development in this deltaic region has caused highly dense areas of human activity and led to serious contamination including heavy metals and persistent organic pollutants (POPs).

Nine sampling sites, namely Barrackpur (S_1), Dakshineswar (S_2), Babughat (S_3), Budge budge (S_4), Ulubaria (S_5), Diamond Harbor (S_6), Frezergunge (S_7), Gangasagar (S_8), and Haribhanga (S_9) were selected considering the existence of typical sediment dispersal patterns along the drainage network systems (as shown in Figure 1) and their position was fixed by a global positioning system (GPS). The stations are representative of the variable environmental and energy regimes that cover a wide range of substrate behavior, wave–tide climate, and intensity of bioturbation (animal–sediment interaction), geomorphological–hydrodynamic regimes and distances from the sea (Bay of Bengal). The sites are exposed to a variable level of heavy metal contamination mainly from anthropogenic sources as mentioned earlier. Six sampling sites (S_1 to S_6) have been chosen along the lower stretch of Hugli River Estuary, while residual three sites (S_7 to S_9) were taken into account in the coastal regions of Sundarban wetland. All sampling sites together with the main stresses to which they are subjected are presented in Table 1.

During winter months (January–March 2009) surface sediment samples weighting 10 g were randomly collected in triplicate from the top 3–5 cm of the surface at each sampling site during low tide using a grab sampler, pooled and thoroughly mixed. Immediately after collection, the samples were placed in sterilized plastic bags in the ice box and transported to the laboratory. Samples were oven dried at 50°C, most gently disaggregated, transferred into precleaned inert polypropylene bags and stored in deep freeze prior to analyses. Each sample was divided into two aliquots: one unsieved (for the determination of sediment quality parameters) and the other sieved through 63 μm metallic sieves (for elemental analyses). Organic carbon content was determined following a rapid titration method [86] and pH with the help of a deluxe pH meter (model no. 101E) using combination glass electrode manufactured by M.S. Electronics Pvt. Ltd. (India). Mechanical analyses of sediment were done by sieving in a Ro-Tap Shaker manufactured by W.S. Tyler Company, Cleveland, Ohio.

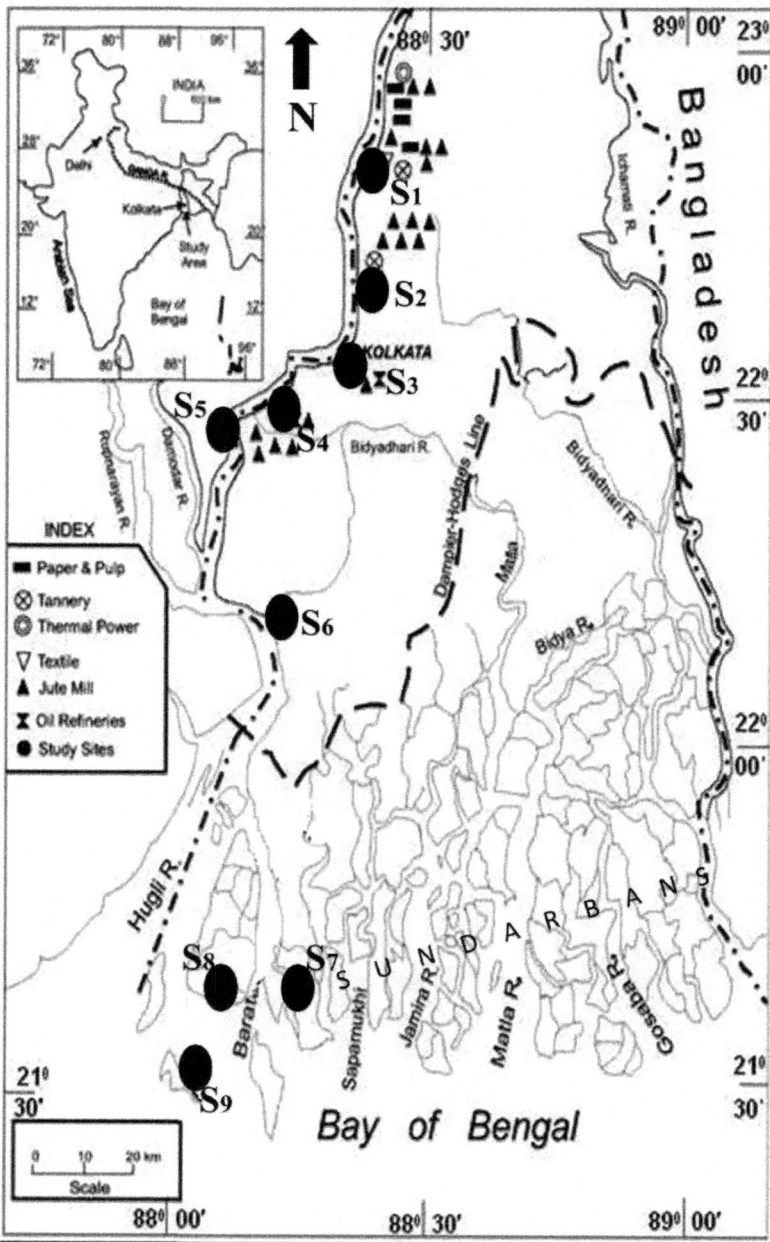

Figure 1. Map showing the location of the monitoring sites (S_1 to S_9) covering Hugli River Estuary and Sundarban mangrove wetland along with the location of the major industries.

Table 1. Details of the nine sampling sites and the main stresses to which they are subjected.

Station number	Site	Main stresses
$s1$	Barrackpur	Industrial effluents, domestic sewage disposal, boating, bathing.
$s2$	Dakshineswar	Industrial and domestic effluents, boating, bathing, idol immersion site.
$s3$	Babughat	Power plant discharges, domestic sewage, boating, idol immersion site.
$s4$	Budge budge	Domestic and industrial effluents, bathing, boating.
$s5$	Ulubaria	Domestic and industrial effluents
$s6$	Diamond Harbour	Boating, recreational activities, bathing, fishing, jetties for fishing trawlers
$s7$	Frezergunge	Tourist activities, ferry services, fishing
$s8$	Gangasagar	Boating, tourist activities, dredging, fishing, agricultural, domestic and aquaculture practices
$s9$	Haribhanga	Boating, fishing and ferrying

Analytical Procedure

To determinate the total element concentration, sediment samples were digested in polytetrafluoroethylene vessels with aqua regia (HCl/HNO$_3$, 3:1) and HF neutralized with H$_3$BO$_3$ in a 650 W microwave oven (CEM MDS 2000) with a program consisting of a 20-min ramp and a 30-min hold at 100% power in pressure and temperature controlled conditions (150 psi and 180°C). The digested samples were filtered, transferred to polyethylene containers and stored at +4°C until analysis. All reagents were Suprapur® grade (Merck). Reagent blank was processed with the samples and did not show any significant contamination. Accuracy of the procedure was checked using two different certified reference materials (CRM): MESS-2 and PACS-2, which are both marine sediments certified by the National Research Council of Canada for the element content. The MESS-2 recovery ranged between 91% and 116% for all the elements (Table 2). Precision, calculated as relative standard deviation (RSD%), resulted always lower than 5%.

Table 2. Results of certified reference materials MESS-2 and PACS-2 as well as the observed values. All the values are expressed in µg/g of dry weight. MESS-2 and PACS-2 recovery rates are also reported.

	Al	As	Cd	Co	Cr	Cu	Fe	Mn	Ni	Pb	Zn
Found MESS-2	86,613± 17,773	24.0± 2.4	0.230± 0.010	15.6± 1.5	112± 12	38.4± 6.1	47,385± 3,668	372± 42	54.1± 3.8	20.0± 1.4	170±12
Found PACS-2	70,190± 3,784	29.2± 1.3	1.59± 0.80	12.8± 0.6	94.9± 4.4	307± 22	46,630± 1,411	465± 23	44.3± 2.5	184± 10	398±16
Certified MESS-2	85,698± 2,600	20.7± 0.8	0.240± 0.010	13.8± 1.4	106± 8	39.3± 2.0	43,504± 2,266	365± 21	49.3± 1.8	21.9± 1.2	172±16
Certified PACS-2	66,125± 3,184	26.2± 1.5	2.11± 0.15	11.5± 0.3	90.7± 4.6	310± 12	43,738± 585	440± 19	39.5± 2.3	183±8	364±23
Recovery MESS-2	101%	116%	95.8%	113%	106%	97.8%	109%	102%	110%	91.2%	99.0%
Recovery PACS-2	106%	111%	75.4%	111%	105%	99.1%	107%	106%	112%	100%	109%

In this study, we followed the sequential extraction procedure proposed by the European Union's Standards, Measurements and Testing program [3]. Selective extraction is based on the procedure used by Tessier et al. [2] with improvements made according to the BCR, which examined and finally eliminated irreproducibility sources. It is made up of three steps, which dissolve the following phases, respectively: exchangeable and bound to carbonate, bound to Fe and Mn oxides and hydroxides, bound to organic matter and sulphides. Exchangeable and bound to carbonate phase (phase 1) is extracted with 0.11 M acetic acid, while the fraction bound to Fe–Mn oxides (phase 2) with 0.5 M hydroxylamine hydrochloride, adjusted to pH 2 with nitric acid (65%). The phase bound to organic and sulphides (phase 3) is extracted with 8.8 M hydrogen peroxide (stabilized at a pH included between 2 and 3), treated at 80°C in a microwave oven using a program consisting of a 30-min ramp and a 60-min hold at 50% power in pressure and temperature controlled conditions (80 psi and 85°C), and 2 M ammonium acetate adjusted to pH 2 with nitric acid (65%). Each extraction was carried out overnight (16 h) at room temperature. All the reagents employed were Tracepur® grade (Merck Eurolab, Italy). After each extraction, the samples were separated from the aqueous phase by centrifuging at 4, 000 rpm for 15 min. The sediments were washed with Milli-Q water and centrifuged again. The wash water was added to supernatants. The element content of the residual phase was obtained from the difference between the total content and the sum of phases 1, 2 and 3, according to Ianni et al. [37, 38], Ramirez et al. [39], and Mester et al. [27]. Sequential extraction reagent blanks showed no detectable contamination. Accuracy of the procedure was checked with BCR-701 (SM&T). The recovery rates for trace elements in the standard reference material ranged between 77% and 118% (Table 3). Precision, calculated as RSD%, resulted generally lower than 5%, except As and Cr in the phase 1 (~20%).

The elemental concentrations were determined with an inductively coupled plasma atomic emission spectrometer Vista Pro (Varian), with the internal standard method. Cadmium was determined by electrothermal atomization atomic absorption spectrometry. A Varian Spectra A300 spectrometer with Zeeman effect background correction and autosampler Varian Model 96 was used employing the standard addition method for calibration. All the metal analyses were performed at the Department of Chemistry and Industrial Chemistry of the University of Genoa (Genoa, Italy).

Table 3. Results of certified reference materials BCR-701 as well as the observed values (expressed in µg/g of dry weight) together with recovery rates for each step. n.a.= not available.

	Al	As	Cd	Co	Cr	Cu	Fe	Mn	Ni	Pb	Zn
Found BCR-701 step 1	198±1	2.57±0.28	6.09±0.09	2.06±0.08	2.41±0.51	47.7±1.7	43.8±5.8	180±1	14.5±0.3	3.38±0.35	185±4
Found BCR-701 step 2	3,451±46	16.5±0.3	3.37±0.08	3.22±0.03	39.2±0.4	100±2	7,042±106	128±3	24.5±0.4	111±2	102±1
Found BCR-701 step 3	1,912±74	3.09±0.20	0.28±0.01	1.86±0.17	169±4	64.8±1.5	1,147±56	31.9±2.6	17.4±1.7	7.15±0.12	58.4±5.0
Certified BCR-701 step 1	n.a.	n.a.	7.34±0.35	n.a.	2.26±0.16	49.3±1.7	n.a.	n.a.	15.4±0.9	3.18±0.21	205±6
Certified BCR-701 step 2	n.a.	n.a.	3.77±0.28	n.a.	45.7±2	124±3	n.a.	n.a.	26.6±1.3	126±3	114±5
Certified BCR-701 step 3	n.a.	n.a.	0.27±0.06	n.a.	143±7	55.2±4.0	n.a.	n.a.	15.3±0.9	9.3±2.0	54.2±2.0
Recovery step 1	n.a.	n.a.	83.0%	n.a.	107%	96.8%	n.a.	n.a.	94.4%	106%	90.3%
Recovery step 2	n.a.	n.a.	89.4%	n.a.	85.9%	80.6%	n.a.	n.a.	92.3%	87.9%	89.4%
Recovery step 3	n.a.	n.a.	104%	n.a.	118%	117%	n.a.	n.a.	114%	76.9%	108%

Statistical Analyses

Principal component analysis (PCA) was used to characterize the metal composition in sediments, and cluster analysis was used for grouping the sampling stations. Principal component analysis (PCA) is a multivariate statistical technique used for data reduction and for deciphering patterns within large sets of data. With PCA, a large data matrix is reduced to two smaller ones that consist of principal component (PC) scores and loadings. PC loadings are eigenvectors of the correlation or covariance matrix depending on which is used for the analysis. The PC scores contain information on all of the variables combined into a single number, with the loadings indicating the relative contribution of each variable to that score [87]. Hierarchical cluster analysis (HCA) characterizes similarities among samples by examining interpoint distances representing all possible sample pairs in high-dimensional space. The sample similarities are represented on two dimensional diagrams call dendrograms [88]. All statistical analyses were performed using the computer software STATISTICA (StatSoft, Inc. 2001).

Results and Discussion

Sediment Geochemistry

Table 4 shows values of pH; organic carbon (%); and percentage of sand, silt, and clay in sediments of the nine sampling sites. Organic carbon values, ranging from 0.22% (in station S_8) to 1.02% (in station S_2), are low in comparison with values found in sediments from other Indian coastal areas, such as Gulf of Mannar [89], Cochin [90], and Muthupet mangroves [91]. The low organic carbon values might be related with the poor absorbability of organics on negatively charged quartz grains, which predominate in sediments in this estuarine environment [92]. In addition, the constant flushing activity by tides along with the impact of waves can support the low percentage of organic carbon in the sediments. The sediments of the studied stations are characterized by slightly basic pH (7.50–8.36) with maximum values recorded in the stations closest to the sea (stations S_6, S_8, and S_9) and minimum in station S_7.

Table 4. Geographical position, physicochemical and textural properties of sediment samples of 9 sampling sites.

Stations	Latitude and Longitude	Salinity	pH	Organic carbon (%)	Sand (%)	Silt (%)	Clay (%)
S1	22°43' 16" N 88°21' 20" E	0	7.86	0.35	4	87.1	8.9
S2	22°39' 17" N 88°12' 25" E	0	7.80	1.02	1	76.5	22.5
S3	22°33' 53" N 88°20' 19" E	0	7.90	0.52	2.24	41.97	55.79
S4	22°30' 10" N 88°11' 48" E	0–2.5	7.60	0.74	18.25	47.42	34.33
S5	22°28' 06" N 88°06' 54" E	0–1	7.90	0.91	16.7	69.6	13.7
S6	22°11' 14" N 88°11' 15" E	0–5.6	8.36	0.56	3.15	41.13	55.71
S7	21°34' 44" N 88°15' 03" E	30–34.3	7.50	0.36	98.02	0.18	0
S8	21°38' 15" N 88°03' 53" E	32–35	8.14	0.22	32.85	58.45	8.7
S9	21°34' 20" N 88°01' 25" E	35	8.10	0.46	39.3	44.25	16.45

These were different from the low pH values in most of the mangrove swamps in Hong Kong [93], where sediments were not frequently flooded by the tide and become acidic in reducible conditions. With respect to texture, the sediment samples show a variable admixture of sand, silt, and clay. Clay fractions dominate in low-energy areas of suspensional deposits. On the contrary, silt, and sand dominates where the energy level is high. Sediments from station S_7 contain higher percentage of sand (98%) compared to the others, while sediments from S_1, S_2, S_5, and S_8 contains higher percentage of silt (more than 50%) compared to the others. A variable mixture of sand, silt, and clay is present in the other stations and reflect a variable amount of erosion and deposition.

Total Element Concentrations

Total element concentrations in the investigated stations varied in a narrow range of values (Table 5)and were comparable with data obtained for other Indian coastal areas [94, 95]. Datta and Subramanian [96] found very similar trace element concentrations throughout the Bengal Basin, where anthropogenic perturbation is low and river channel may receive a several centimeter-thick sediment layer in a single event during peak flow, preventing to bear the signature of an accumulation of trace elements. The highest concentrations for As, Cu, Fe, Mn, and Ni were measured at station S_9 while for Cd and Pb at station S_3, close to Calcutta city (about 4.5 million residents, but about 14.2 million including suburbs). An anthropogenic input from vehicular traffic and in-dustrial activities may cause high Cd and Pb con-centrations measured in samples collected in the Calcutta urban area. The lowest element concentrations were found at station S_5. Very low (close to the detection limit) Cd concentration was found in the coastal stations (S_7, S_8, and S_9).

The geoaccumulation index (I_{geo}) of Muller [97] has been calculated for the analyzed elements, by comparing current concentrations with pre-industrial levels, in order to estimate the metal contamination in sediments. The equation used for the calculation of I_{geo} is: $\log_2 (C_n/1.5\ B_n)$, where C_n is the measured content of element "n" and B_n the element's content in "average shale" [98]. Factor 1.5 is used because of possible variations in background values for a given element in the environment, as well as very small anthropogenic influences [99]. As shown in Figure 2, all sediments fall in class 0 for Al, Co, Cr, Cu, Fe, Mn, Ni, Pb, and Zn, therefore the area is not contaminated for these elements. Unlike the Hugli river, in other rivers of the Bengal Basin, such as Meghna and Brahmaputra, Cr exhibits higher I_{geo} values respect to the other elements [96]. For Cd, two stations fall in class 1 and three in class 2 exhibiting a moderate contamination for this element. In all stations, As falls in class 2 (moderate pollution). In this area, As contamination was already observed in previous studies and it is probably due to groundwater contamination [100]. This contamination can have natural origin, such as coal seams in Rajmahal basin and arsenic mineral in mineral rocks in the upper reaches of the Ganges river system. The highly reducing nature of groundwater would reduce As, causing the possible desorption of As [101].

Table 5. Total element concentrations (μg/g) in sediments of 9 sampling sites (instrumental precision, calculated as RSD%, resulted lower than 5% for each element in all samples).

Stations	Al	As	Cd	Co	Cr	Cu	Fe	Mn	Ni	Pb	Zn
S1	70,289	8.81	0.165	13.0	67.6	27.8	37,737	591	31.9	20.4	86.6
S2	70,879	8.44	0.452	14.0	74.8	36.8	39,405	625	34.2	22.3	90.7
S3	72,134	8.65	1.79	14.5	73.5	32.3	40,070	712	35.0	33.2	83.1
S4	72,613	8.49	0.492	14.9	76.8	27.9	40,303	726	35.0	19.6	80.4
S5	62,044	6.41	0.220	12.1	58.2	21.1	33,428	597	27.5	17.0	64.1
S6	64,325	6.79	0.106	12.0	64.8	32.0	34,273	613	31.3	17.9	69.6
S7	77,529	7.77	0.044	14.0	75.1	28.4	40,084	389	38.1	20.5	74.4
S8	68,146	8.08	0.027	12.7	62.5	22.2	36,786	511	34.3	19.7	61.4
S9	72,666	9.40	0.044	14.1	74.2	36.6	40,838	785	40.1	22.9	74.9

Speciation Patterns

The potential environmental risk of trace elements in sediments is associated with both their total content and their speciation. The chemical partitioning of the considered elements (Al, As, Cd, Co, Cr, Cu, Fe, Mn, Ni, Pb, and Zn) from each extraction step has been described. Aluminum, Cr, and Fe are present mainly in the residual phase, representing 95.8–96.8%, 88.9–91%, and 83.0–94.7% of the total concentration, respectively, which implies that these elements are strongly linked to the inert fraction of the sediments. This result was in good agreement with data reported by several studies carried out worldwide in marine coastal areas [45, 46, 78, 102]. The high percentage of Fe in the residual phase indicates that most of the Fe exists as crystalline Fe peroxides (goethite, limonite, magnetite, etc.). The remaining Fe is associated with the reducible phase (mean, 11.25%). Large amounts of Fe accumulate in the residual phase probably because it is basically of natural origin (it is the most common element in the earth's crust).

Concentrations of Al, Fe, and Cr are very low in exchangeable phase (0.08%, 0.26%, and 1.72% as mean values, respectively), limiting their potential toxicity as pollutants. It should be noted that sediments always act as reservoir for elements; therefore, their potential risk of pollution to environment has always to be considered.

Arsenic, Co, Ni, and to some extent Zn, are found mainly in the residue (~50% of the total concentration). Nickel and Co are associated to the residue respectively for 56% and 74% of the total concentration, with a speciation similar in all the samples. A mean of 23% of Co is present in the phase 2. The highest percentage of labile Co (~13%) was measured in S_6 (Diamond Harbour) and S_8(Gangasagar) and can be due to a recent input of this element. The dominant proportion of Ni in the residual phase is in agreement with the results of other studies [27, 46]. Nickel is present, apart from the residue, in phases 2 and 3 (about 10% in each phase). Arsenic is distributed mainly between the residual (mean 47%), the reducible, and the oxidizable phases (mean 19% and 22%, respectively). Acharyya et al. [101] observed that As is adsorbed to iron-hydroxide-coated sand grains and to clay minerals in the sediments of the Ganges delta from West Bengal. Among the studied elements, As is found with the greatest proportion in the oxidizable phase coinciding with organic and sulfur compounds. Arsenic is present in the phase 1 for about 10% of the total content, in station S_7 phase 1 percentage rises up to 16%. The lower land alluvial basin of the Ganges River is recognized as an arsenic-affected area. Arsenic in solution probably is easily entrapped in the fine grained organic-rich sediments deposited in the Ganges delta [101]. The percentage of silt (lower than 70% except in S_1 and S_2) may have contributed to a low retention of dissolved As since coarse sediments are less efficient at retaining As.

Cadmium was mainly present in the labile phase (more than 60%) in all the stations with the exception of station S_7, where the Cd labile percentage represents only 25% of the total concentration. Cadmium concentrations were negligible in phases 2 and 3. The highest labile Cd concentration was measured at station S_3, the closest to the city of Calcutta. Datta and Subramanian [96] found that the concentrations of elements in the non-detrital phases were higher in stations sampled in the Hugli river around Calcutta than in samples collected along Brahmaputra and Meghna rivers. The petroleum refinery, industrial, and mining effluents carried by the Hugli river may be responsible for this higher concentrations of non detrital fractions.

About 40% of the total Cu concentration is associated to the residue, while 33% of Cu is bound to Fe-Mn oxide and hydroxide (phase 2). The high percentage of Cu in the residue is likely due to the fact that Cu is easily chemisorbed on or incorporated in clay minerals [103]. All the samples showed lower Cu concentrations in exchangeable phase, with percentage ranging from 7% (S_7) to 22% (S_5), with a mean of 15%. Copper is characterized by high complex constant with organic matter thus it can be hypothesized that Cu is bound to labile organic matter such as lipids, proteins, and carbohydrates. On the other hand, high-element concentration in labile phase could be related to recent coastal input [39].

Manganese was found in all the four sediment phases, as observed by other researchers [45, 104]. Manganese is the most mobile element since it is present with the highest percentage (a mean of 42%) in the labile phase. This is probably because of the known close association of Mn with carbonates [105] as endorsed by other workers [69, 106]. In this phase, weakly sorbed Mn retained on sediment surface by relatively weak electrostatic interactions may be released by ion exchange processes and dissociation of Mn-carbonate phase [2]. The result indicates that considerable amount of Mn may be released into environment if conditions become more acidic [107]. The highest Mn labile percentage was measured in S_6 (57%). Differently, in S_7, Mn in the residue represents 65% of the total concentration, while the labile Mn is only 15%. A substantial Mn percentage was also found in the residue (mean 37.8%), followed by the reducible phase (14.7%), in which Mn exists as oxides and may be released if the sediment is subjected to more reducing conditions [108].

The major geochemical phase for Pb in these sediments was the Fe-Mn oxides phase (mean 55.7%) followed by the residual phase (mean 30.2%) while lower percentage of the total Pb are bound to exchangeable-labile (mean 5.3%) and oxidizable phases (mean 6.8%). At S_3 (Babughat), the reducible part is as high as 65% and only 19.9% of the total is associated with the residue. Atmospheric input as fallout from vehicular emission can be probably the major input of Pb for this station. The relatively high percentage of Pb in reducible phase is in agreement with the known ability of amorphous Fe–Mn oxides to scavenge Pb from solution [109, 110]. Caille et al. [111] observed that resuspension of anoxic sediment results in a rapid desorption of Pb and Cu adsorbed to sulphides. Thus, a high element percentage in the reducible fraction is a hazard for the aquatic environment because Fe and Mn species can be reduced into the porewaters during early diagenesis by microbially mediated redox reactions [112]. Dissolution will also release Pb associated with oxide phases to the porewater possibly to the overlying water column [113] and to benthic biota [79]. The major sources of Pb are from intensive human activities, including agriculture in the drainage basin [114], auto exhaust emission together with atmospheric deposition [115]. In addition, a substantial contribution from the factories located in the upstream of the Hugli river dealing with Pb producing lead ingots and lead alloys play a vital role as referred by Sarkar et al. [116].

The percentage of Zn in residue is highly variable (38.5–70%) and the distribution pattern in each fraction showed the following order: residual>reducible>oxidizable> exchangeable and bound to carbonates. There was some difference in Zn speciation among the sampling sites: in stations S_1, S_2, and S_3 about 40% of Zn is present in the residue, while in the other stations this percentage increases to more than 60%. In station S_1, the exchangeable and oxidisable phases shared over 22% of the total Zn, whereas labile Zn was as low as 4.6% at S_7. A major part of Zn (16.3%) is associated with Fe–Mn oxide phase, because of the high stability constants of Zn oxides. Iron oxides adsorb considerable quantities of Zn and these oxides may also occlude Zn in the lattice structures [117].

The BCR procedure as discussed above showed satisfactory recoveries, detection limits, and standard deviations for determinations of heavy metals/ metalloid in the sediments. It is evident from the present results of the fractionation studies that the metals/metalloids in the sediments are bound to different fractions with different strengths leading to variations in mobility and availability and some of them show significant spatial variations subject to diverse environmental stresses. This type of association between metals and the sediments can be understood in detail by sequential extraction techniques. Hence the application of sequential extraction is fully justified as the quantification of different forms of metal is more meaningful than the estimation of its total metal concentrations. The strength values can, therefore, give a clear indication of sediment reactivity, which in turn assess the risk connected with the presence of metals in this wetland environment. The results obtained suggest the need for corrective remediation measures due to the higher accumulation of potentially dangerous metals/metalloids, which in most cases exceed the limits established by certain legislation.

Comparison with Sediment Quality Guidelines

Results obtained after total and sequential extraction are compared with Sediment Quality Guidelines (SQGs). Table 6 reports consensus-based values, such as TEC (concentration below which harmful effects on sediment-dwelling organisms were not expected) and PEC (concentration above which harmful effects on sediment-dwelling organisms were expected to occur frequently), and effect range-low and range-medium, such as ERL (concentrations below which adverse biological effects were observed in less than 10% of studies) and ERM (concentrations above which effects were more frequently observed in more than 75% of studies).

Comparing our results with the SQGs, it is revealed that for Pb and Zn in all the stations the measured concentrations are lower than both TEC and ERL. As regards Cd, concentration measured in station S_3 is higher than TEC and ERL but lower than PEC and ERM both in term of total and labile concentration. For this station, some possible toxic effect on benthic organism can be hypothesized, in particular because of the large amount of element bound to the most labile phase of the sediment. Considering Cu, some stations (S_2, S_3, S_6, and S_9) exhibit total concentrations higher than TEC but lower than PEC. Concentrations of Cu are higher than ERL but lower than ERM only in stations S_2 and S_9. Since only 7–22% of total Cu is bound to the labile phase, in all stations Cu labile concentrations are lower than TEC and ERL. Total As concentrations in stations S_1, S_2, S_3, S_4, and S_9 are higher than ERL value but lower than TEC value. Since more than 50% of total As is not found in the residue, attention should be paid to a change in the environment conditions which could induce a release of As from the sediments. Total Ni and Cr concentrations are higher than TEC (Ni is also higher than ERL) but lower than PEC (and ERM in the case of Ni) in all the stations. Nevertheless, more than 70% of Ni as well as 90% of Cr are present in the residual fractions, therefore adverse impacts on organisms is very much negligible.

Table 6. Sediment Quality Guidelines concentrations with respect to total and labile element concentrations found in the analyzed samples (expressed as µg/g of dry weight).

Element	Phase	Si<TEC	TEC	Si<TEC<PEC	PEC	Si<ERL	ERL	ERL<Si<ERM	ERM
As	Total	All	9.79	None	33	S5,S6,S7,S8	8.2	S1,S2,S3,S4,S9	70
	Labile	All		None		All		None	
Cd	Total	S1,S2,S4,S5,S6,S7,S8,S9	0.99	S3	4.98	S1,S2,S4,S5,S6,S7,S8,S9	1.2	S3	9.6
	Labile	S1,S2,S4,S5,S6,S7,S8,S9		S3		S1,S2,S4,S5,S6,S7,S8,S9		S3	
Cr	Total	None	43.4	All	111	All	81	None	370
	Labile	All		None		All		None	
Cu	Total	S1,S4,S5,S7,S8	31.6	S2,S3,S6,S9	149	S1,S3,S4,S5,S6,S7,S8	34	S2,S9	270
	Labile	All		None		All		None	
Ni	Total	None	22.7	All	48.6	None	20.9	All	51.6
	Labile	All		None		All		None	
Pb	Total	All	35.8	None	128	All	46.7	None	218
	Labile	All		None		All		None	
Zn	Total	All	121	None	459	All	150	None	410
	Labile	All		None		All		None	

Mean sediment quality guidelines quotients (mSQGQ) have been developed for assessing the potential effects of contaminant mixtures in sediments [118]: they are determined by calculating the arithmetic mean of the quotients derived by dividing the concentrations of chemicals in sediments by their respective SQGs. The probability of observing sediment toxicity can be estimated by comparing the mSQGQ in a sample to previously published probability tables. It is important to keep in mind that mSQGQs cannot be used to accurately predict the uptake and bioaccumulation of sediment-bound chemicals by fish, wildlife, and humans, even if there is considerable evidence that this assessment tool can be predictive of the presence or absence of toxic effects [118].

SQGQs are calculated for seven elements considering ERM as sediment quality guidelines (Table 7). The mean quotient values ranges from 0.16 in station S_5 to 0.24 in station S_3. Using PEC values instead of ERM, the mean SQGQ ranges from 0.25 in station S_5 to 0.38 in station S_3 (Table 7).

Table 7. Mean Sediment Quality Guidelines Quotients calculated for the nine stations using PEC and ERM as SQGs.

Stations	$SQG_{OPE}C$	$SQG_{OER}M$
S1	0.30	0.19
S2	0.33	0.21
S3	0.38	0.24
S4	0.33	0.21
S5	0.25	0.16
S6	0.28	0.18
S7	0.32	0.21
S8	0.28	0.18
S9	0.34	0.22

Compiled data from multiple data sets reporting 10-day toxicity test conducted on amphipod species in saltwater showed that the incidence of toxicity for a range of SQGQ of 0.25–0.5 is ~35%, while for a mean SQGQ range from 0.1 to 0.25, the incidence of toxicity lowers to ~20%. Measures recorded in a survey of Biscayne Bay (port of Miami and the adjoining saltwater reaches of the lower Miami River, FL, USA) showed that the average amphipod survival (*Ampelisca abdita*) decreased slightly from the least contaminated (ERMQ <0.03) to the intermediate category, (ERMQ included in 0.03–0.2 range) then decreased greatly in the most contaminated sediments (ERMQ included in 0.2–2 range). Therefore, we can presume a low toxicity of sediments sampled in the nine stations for benthic organisms. It is important to note that the benthic response to contaminants covaried among stations with both the mean ERM quotients and the effect of natural factors, such as the sediment texture, TOC, and salinity [118].

Statistical Analyses

The relationships between variables and the differences between stations were evaluated by PCA. The analysis was performed on 36 objects (four sediment phases for nine stations) and 11 variables (Al, As, Cd, Co, Cr, Cu, Fe, Mn, Ni, Pb, Zn). Two significant components were identified explaining 68.3% and 14.5% of the total variance, respectively. By studying the loadings of the variables (Figure 2a) on the components it can be seen that all the elements except Cd, Mn, and Pb are significantly correlated.

Unlike the other elements, most of Cd and Mn is present in the first phase: labile Cd and Mn represent more than 60% and 40% of the total concentration, respectively, except in station S_7. Cadmium and Mn speciation can be ascribed to their considerable affinity for carbonates. Lead is the only element which is bound to the reducible phase for more than 50%. Lead is a very reactive element in water column and, having scavenging type behavior, is easily bound to hydroxy- and oxyligands. Copper is positively and significantly correlated with all elements except Cd and Mn, but with lower correlation coefficients (0.66–0.81).

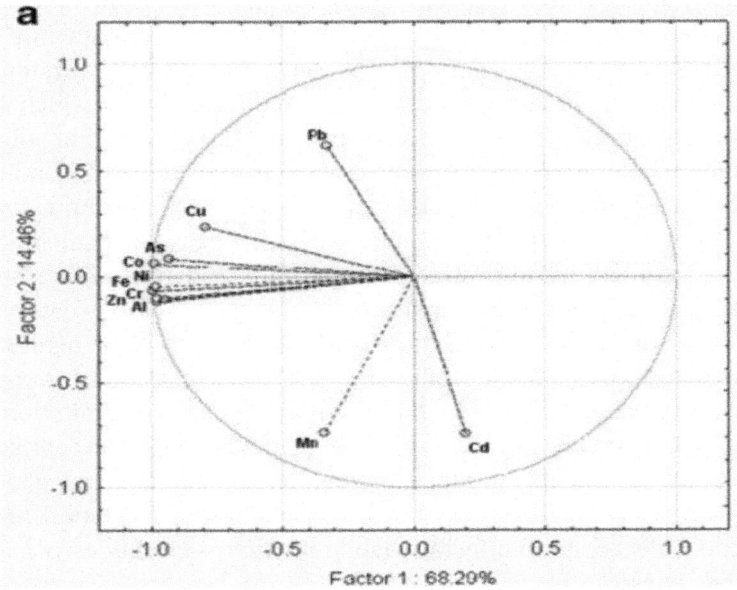

b

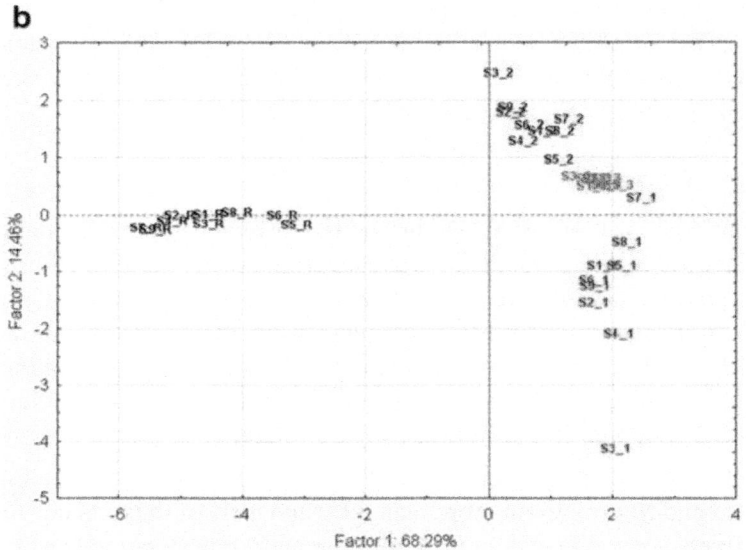

Figure 2. Principal component analysis: a) variable plot; b) score plot (phase 1, 2, 3 and 4, corresponding to labile, reducible, oxidizable and residue, are identifiable by 1, 2, 3 and 4 suffix and different colors in the score plot).

In the score plot (Figure 2b) phases 1, 2, 3, and 4 (corresponding to labile, reducible, oxidizable, and residual phases respectively) are identifiable by 1, 2, 3, and 4 suffix, respectively. In all stations, residue concentrations were characterized by negative values of PC1 and consequently by high concentrations of Al, As, Co, Cr, Fe, Ni, and Zn. Conversely, in the positive PC1 semi-axis labile and oxidizable metal concentrations, which represent a small percentage of the total elements, are distributed. For all stations, reducible concentrations are distributed along the positive PC2 semi-axis, i.e., high Pb concentrations, with a maximum for station S_3 and a minimum for S_5. The group formed by elements bound to organic matter and sulphides (phase 3) is characterized by low values of both PC1 and PC2. Therefore, a low percentage of elements (higher than 20% exclusively for all As data and for Zn in stations S_1 and S_3) is bound to the oxidizable phase, suggesting the presence of an oxidant environment. High Mn and Cd concentrations are associated with negative values of PC2, therefore a relatively high concentration of labile Mn and Cd is present in all samples (in particular in S_3), except station S_7. Samples are prevalently grouped in relation to the sediment geochemical phase, suggesting a similar element speciation among the stations. Station S_7 represents an exception, in fact the labile fraction is closely associated to the oxidizable phase group.

A HCA was carried out by applying Euclidean distances to quantitatively identify specific groups of similar stations. In the dendrogram of the sampling stations (Figure 3), we can note two main clusters: the first represented by station S_7, characterized by the highest element percentage bound to residue, and the second constituted by all the remainder stations. In the second group, a subgroup formed by station S_5 and S_6 can be individuated.

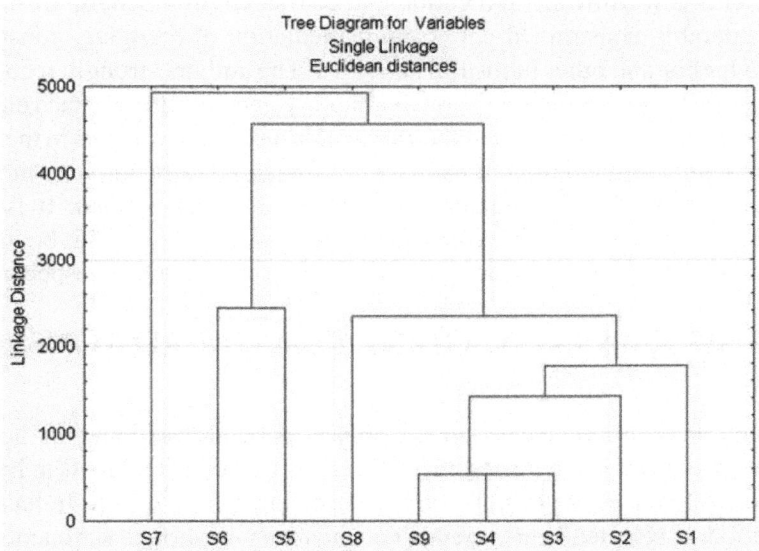

Figure 3. Dendrogram indicating linkage of sites on the basis of element concentrations.

Station S_7 was sampled in a marine coastal environment; it is characterized by a peculiar grain size percentage respect to the other stations, being the sand percentage as high as 98.6%. In general, the concentrations of elements are much higher in fine than in coarse fraction because the fine fraction larger specific surface facilitates absorption processes. As previously noted by Ramirez et al. [39], this pattern is particularly evident for Cd. It is interesting to note that the marine coastal stations S_8 and S_9 are more similar to river stations than to station S_7. Both the stations S_8 and S_9 are, in fact, located in front of the Hugli river runoff, while station S_7 is located easternmost and probably is less influenced by the Hugli river discharge.

Conclusion and Recommendation

The study provides valuable information on the potential mobility of trace elements in sediments collected along the stretch of Hugli River and in the

Sundarban mangrove wetland (northeastern part of the Bay of Bengal). The results obtained adopting BCR sequential extraction method provided the following important information: (i) Al, Cr, and Fe were found mostly in the residual phase while the other elements were found in the four phases sharing different proportions; (ii) the dominant Cd, Mn, and Pb proportion was found in the non-residual fractions and (iii) Mn had the highest percentage in the labile phase. This is worthwhile to mention that coastal environment of West Bengal is considerably constrained due to implementation of dredging, construction of port/ harbor and other industrial activities. The authors strongly recommend for periodical monitoring on the bioavailability and mobility of trace elements, control the mixing of effluent of the concentration of heavy metals in the region, environmental remediation, treatment of industrial effluent and municipal wastewater for effective management of this estuarine system. It is wisely suggested that an environmental recovery framework should be urgently implemented to avoid extension of heavy metal contamination (especially As).

CASE STUDIES FROM OTHER COASTAL REGIONS IN INDIA

Although the importance of metal speciation and fractionation has been realized in developed countries, the subject has not really taken off in India and only few references are available on the speciation of metals in Indian rivers. Speciation of selected heavy metals geochemistry in surface sediments (n=10 was studied by Venkatramanan et. al. [119] from Tirumalairajan river estuary, east coast of India. The results obtained from sequential extraction showed that a larger portion of the metals were associated with the residual phase, although they are available in other fractions too.

Trace metal fractionation in the Pichavaram mangrove–estuarine sediments in southeast coast of India was studied by Ranjan et. al. [120] considering the pronounced changes due to occurrence of tsunami (2004). A 5-step sequential extraction procedure was applied to assess the effects of tsunami on mobility and redistribution of selected elements (Cd, Cr, Cu, Fe, Mn, Ni, Pb, and Zn) in coastal sediments revealed that metals in the residual fraction (lattice bound) had the highest concentration suggesting their non-availability and limited biological uptake in the system. Majority of the metals (except Mn) do not constitute a risk based on the different geochemical indices.

Fractionation of selected metals in the sediments of Cochin estuary and Periyar River (southwest coast of India) was studied by Mohan et. al. [121]. The results reveal that remobilization potential of metals bound is in the range of low to medium risk to various sedimentary phases is different and is based on bond strength. Therefore, the strength values can give a clear indication of sediment reactivity that can be used to assess the risk related with metals to the aquatic organisms.

RISK ASSESSMENT CODE (RAC)

The risk assessment code (RAC) mainly applies the sum of the exchangeable and carbonate bound fractions for assessing the availability of metals in sediments. These fractions are considered to be weakly bonded metals which may equilibrate with the aqueous phase and thus become more rapidly bioavailable [11, 33]. This is important because the fractions introduced by anthropogenic activities, such as agricultural runoff and tourism, are typified by the adsorptive, exchangeable, and bound to carbonate fractions, which are weakly bonded metals that could equilibrate with the aqueous phase and thus become more rapidly bioavailable [122]. According to RAC guideline (Table 8), for any metal, soil/sediment which can release in exchangeable and carbonate fractions, less than 1% of the total metal will be considered safe for the environment and soil/sediment with 11-30% carbonate and exchangeable fractions will be at medium risk to the environment. On the contrary, soil/ sediment releasing in the above fractions more than 50% of the total metal has considered being highly dangerous, which can be easily enter the food chain [123].

Table 8. Criteria of Risk Assessment Code [123].

Grade	Exchangeable and bounded to carbonate metal (%)	Risk
I	<1	No risk
II	1 – 10	Low risk
III	11 – 30	Medium risk
IV	31 – 50	High risk
V	>50	Very high risk

Heavy-Metal Fractionation in surface sediments was studied by Dhanakumar et. al. [124] in the Cauvery river estuarine region, southeastern coast of India. The results revealed that most of the samples fall under the category from low- to high-risk class and from low-risk to very high-risk class in terms of labile fractions of Pb as well as Zn and Cu, respectively.

CONCLUSION

From the above discussion it is revealed that geochemical fractionation approach to the chemical speciation has provided a useful tool and opens a new dimension in assessing the potential mobility/bioavailability of heavy metals and metalloids in soils/sediments and opens a new dimension in the field of ecology and environmental chemistry. More efficient, non-laborious and time saving processes techniques in this field of chemical speciation are also coming up to get valid information regarding geochemical behavior of soils/sediments. Besides geochemical fractionation, Dezileau et al. [125] opined that total Fe or Fe/Al may be used to infer millennial-scales climate changes in the south eastern pacific while performing sequential extraction of Fe in marine sediments from the Chileau continental margin. However, the chemical partitioning should be carefully used in the assessment of environmental pollution as large amount of metals may naturally occur as anthropogenic fractions (including loosely bonded ions, sulfide ions and metals associated with sediments).

ACKNOWLEDGEMENTS

The authors gratefully acknowledge full support and cooperation of the Springer press, UK for extending permission in publishing the research paper of the journal Environmental Monitoring & Assessment, vol. 184(12), pp:7561-77, 2012.

This study was partially supported by the European Fund for Economic and Regional Development (FEDER) through the Program Operational Factors of Competitiveness (COMPETE) and National Funds through the Portuguese Foundation for Science and Technology (PEST-C/MAR/UI 0284/2011, FCOMP 01 0124 FEDER 022689).

REFERENCES

1. Gleyzes C, Tellier S, Astruc M. Fractionation studies of trace elements in contaminated soils and sediments: a review of sequential extraction procedures. Trends in Analytical Chemistry 2002; 21 (6) 451-467.

2. Tessier A, Campbell PGC, Bisson M. Sequential extraction procedure for the speciation of particulate trace metals. Analytical Chemistry 1979; 51 844-851.

3. Sahuquillo A, Rigol A, Rauret G. Overview of the use of leaching/extraction tests for risk assessment of trace metals in contaminated soils and sediments. Trends in Analytical Chemistry 2003; 22(3) 152-159.

4. Calmano W, Hong J, Förstner U. Binding and mobilization of heavy metals in contaminated sediments affected by pH and redox potential. Water Science and Technology 1993; 28(8/9) 53-58.

5. Eggleton J, Thomas KV. A review of factors affecting the release and bioavailability of contaminants during sediment disturbance events. Environment International 2004; 30 973-980.

6. Ure AM, Quevauviller P, Muntau H, Griepink B. Speciation of heavy metals in solids and harmonization of extraction techniques undertaken under the auspices of the BCR of the Commission of the European Communities. International Journal of Environmental Analytical Chemistry 1993; 51 135-151.

7. Florence TM. The speciation of trace elements in waters. Talanta 1982; 29 345-69.

8. Lung W. Speciation analysis – why and how? Fresenius' Journal of Analytical Chemistry 1990; 337 557-564.

9. Schroeder WH. Development in the speciation of mercury in natural waters. Trends in Analytical Chemistry 1989; 8 339-342.

10. Rauret G, Rubio R, Lopez-Sanchez JF, Cassassas E. Determination and speciation of copper and lead in sediments of a Mediterranean River (River Tenes, Catalonia, Spain). Water Research 1988; 22(4) 449-51.

11. Pardo R, Barrado E, Perez L, Vega M. Determination and association of heavy metals in sediments of the Pisuerga River. Water Research 1990; 24(3) 373-379.

12. Jardo CP, Nickless G. Chemical association of Zn, Cd, Pb and Cu in soils and sediments determined by the sequential extraction technique. Environmental science & Technology Letters 1989; 10 743-52.

13. Tessier A, Campbell PGC, Bisson M. Heavy metal speciation in the Yamaska and St. Francois rivers (Quebec). Canadian Journal of Earth Sciences 1980; 17 90-105.

14. Elsokkary LH, Muller G. Assessment and speciation of chromium, nickel, lead and chromium in the sediments of the river Nile, Egypt. Science of the Total Environment 1990; 97/98 455-63.

15. Ure AM. Single extraction schemes for soil analysis and related applications. Science of the Total Environment 1996; 178(1/3) 3-10.

16. Rauret G. Extraction procedures for the determination of heavy metals in contaminated soil and sediment. Talanta 1998; 46(3) 449-455.

17. Quevauviller PH, Lachica M, Barahona E, Rauret G, Ure A, Gomez A, Muntau H. Interlaboratory comparision of EDTA and DTPA procedures prior to certification of extractable trace elements in calcareous soil. Science of the Total Environment 1996; 178(1/3) 127-132.

18. Quevauviller PH, van der Slootr HA, Ure A, Muntau H, Gomez A, Rauret G. Conclusions of the workshop: harmonization of leaching/ extraction tests for environmental risk assessment. Science of the Total Environment 1996; 178(1/3) 133-139.

19. Bacon JR, Davidson CM. Is there a future for Sequential chemical extraction? Analyst 2008; 133 25-46

20. Förstner U, Lechsber R, Davis RA, Hermitte PL. (eds.), Chemical methods for assessing bioavailable Metals in Sludges. Elsevier, London, 1985.

21. Meguellati M, Robbe D, Marchandise P, Astruc M, Proceedings International Conference on Heavy Metals in the Environment, Heidelberg CEP Consultants, Edinburgh, 1983, p. 1090.

22. Filgueiras AV, Lavilla I, Bendicho C. Chemextraction for metal partitioning in environmental solid samples, Journal of Environmental Monitoring 2002; 4 832-857.

23. Lopez-sanchez JF, Rubio R, Rauret G. Comparison of two sequential extraction procedures for trace metal partitioning in sediments. International Journal of Environmental Analytical Chemistry 1993; 51 113-121.

24. Hlavay J, Polyak K. Chemical speciation of elements in sediment samples collected at Lake Balaton. Microchemical Journal 1998; 58 281-290.

25. Tokalioglu S, Kartal S, Elçi L. Determination of heavy metals and their speciation in lake sediments by flame atomic absorption spectrometry after a four-stage sequential extraction procedure, Analytical Chimica Acta 2000; 413 33-40.

26. Belazi AU, Davidson CM, Keating GE, Littlejohn D. Determination and speciation of heavy metals in sediments from the Cumbrian coast, NW England, UK. Journal of Analytical Atomic Spectrometry 1995; 10 233-240.

27. Mester Z, Cremisini C, Ghiara E, Morabito R. Comparison od two sequential extraction procedures for metal fractionation in sediment samples. Analytical Chimica Acta 1998; 259 133-142.

28. Gomez-Ariza JL, Giraldez I, Sanchez-Rodas D, Morales E. Comparison of the feasibility of three extraction procedures for trace metal partitioning in sediments from southwest Spain, Science of Total Environment 2000; 246 271-283.

29. Zhang T, Shan X, Li F. Comparison of two sequential extraction procedures for speciation analysis of metals in soils and plant availability. Communications in Soil Science and Plant Analysis 1998; 29 1023-1034.

30. Albores AF, Cid BP, Gomez P, Lopez EF. Comparison between sequential extraction procedures and single extrations for metal partitioning in sewage sludge samples. Analyst 2000; 125 1353-1357.

31. Ho MD, Evans GJ. Operational speciation of cadmium, copper, lead and zinc in the NIST standard reference materials 2710 nad 2711 (Montana soil) by the BCR sequential extraction procedure and flame atomic spectrometry. Analytical Communications 1997; 34 353-364.

32. Zemberyova M, Bartekov J, Hagarov I. The utilization of modified BCR three-step sequential extration procedure for the fractionation of Cd, Cr, Cu, Ni, Pb and Zn in soil reference materials of different origins, Talanta 2006; 70 973-978.

33. Singh KP, Mohan D, Singh VK, Malik A. Studies on distribution and fractionation of heavy metals in Gomti river sediments- a tributary of the Ganges India. Journal of Hydrology 2005; 312 14-27.

34. Davidson CM, Duncan AL, Littlejohn D, Ure AM, Garden LM. A critical evaluation of the three-stage BCR sequential extraction procedure to assess the potential mobility and toxicity of heavy metals in industrially-contaminated land. Analytica Chimica Acta 1998; 363(1) 45-55.

35. McLaren RG, Crawford D. Studies on soil copper I. The fractionation of copper in soils. Journal of Soil Science 1973; 24(2), 172-181.

36. Kersten M, Förstner U. Speciation of trace elements in of water-soluble organic components. sediments and combustion waste.In Ure AM, Davidson CM (ed.) Chemical speciation in the environment. Blackie Academic and Professional, Glasgow, UK. 1995; 234-275.

37. Ianni C, Magi E, Rivaro P, Ruggieri N. Trace metals in Adriatic coastal sediments: distribution and speciation pattern. Toxicological and Environmental Chemistry 2000; 78 73-92.

38. Ianni C, Ruggieri N, Rivaro P, Frache R. Evaluation and comparison of two selective extraction procedures for heavy metal speciation in sediments. Analytical Sciences 2001; 17 1273-1278.

39. Ramirez M, Massolo S, Frache R, Correa J. Metal speciation and environmental impact on sandy beaches due to El Salvador copper mine, Chile. Marine Pollution Bulletin 2005; 50 62-72.

40. Borovec Z, Tolar V, Mraz L. Distribution of some metals in sediments of the central part of the Labe (Elbe) River, Czech Republic. Ambio 1993; 22 200-205.

41. Campanella L, D'Orazio D, Petronio BM, Pietrantonio E. Proposal for a metal speciation study in sediments. Analytica Chimica Acta 1995; 309 387-393.

42. Zdenek B. Evaluation of the concentrations of trace elements in stream sediments by factor and cluster analysis and the sequential extraction procedure. Science of the Total Environment 1996; 177 237-250.

43. Gomez-Ariza JL, Giraldez I, Sanchez-Rodas D, Moralesm E. Metal sequential extraction procedure optimized for heavily polluted and iron Oxide rich sediments. Analytica Chimica Acta 2000; 414 151-164.

44. Lopez-Sanchez JF, Sahuquillo A, Fiedler HD, Rubio R, Rauret G, Muntau H, Quevauviller P. CRM 601, a stable material for its extractable content of heavy metals. Analyst 1998; 123 1675-1677.

45. Usero J, Gamero M, Morillo J, Gracia I. Comparative study of three sequential extraction procedures for metals in marine sediments. Environment International 1998; 24 478-496.

46. Martin R, Sanchez DM, Gutierrez AM. Sequential extraction of U, Th, Ce, La and some heavy metals in sediments from Ortigas River, Spain. Talanta 1998; 46 1115-1121.

47. Agnieszka S, Wieslaw Z. Application of sequential extraction and the ICPAES method for study of the partitioning of metals in fly ashes. Microchemical Journal 2002; 72 9-16.

48. Templeton DM, Ariese F, Cornels R. IUPAC guidelines for terms related to chemical Speciation and Fractionation of elements. Pure and Applied Chemistry 2001; 72 1453-1470.

49. Bordas F, Bourg ACM. A critical evaluation of sample for storage of Contaminated Sediments to Be investigated for the potential mobility of their heavy metal load. Water. Air. Soil. Pollution 1998; 103 137-149.

50. Kersten M, Frostner U. Chemical fractionation of heavy metals in anoxic estuarine and coastal sediments. Water Science and Technology 1986; 18 121-130.

51. Dold B. Speciation of the most soluble phases in a sequential extraction procedure adapted for geochemical studies of copper sulfide mine waste. Journal of Geochemical Exploration 2003; 80 55-68.

52. Favas PJC, Pratas J, Gomes MEP, Cala V. Selective chemical extraction of heavy metals in tailings and soils contaminated by mining activity: Environmental implications. Journal of Geochemical Exploration 2011; 111 160-171.

53. Pardo R, Vega M, Debán L, Cazurro C, Carretero C. Modelling of chemical fractionation patterns of metals in soils by two-way and three-way principal component analysis. Analytica Chimica Acta 2008; 606 26-36.

54. Cuong TD, Obbard JP. Metal speciation in coastal marine sediments from Singapore from Singapore using a modified BCR-sequential extraction procedure. Applied Geochemistry 2006; 21 1335-1346.

55. Dapaah RK, Takano N, Ayame A. Solvent extraction of Pb (II) from acid medium with zinc Hexamethylenedithiocarbamate followed by back-extraction and subsequent determination by FAAS. Analytica Chimica Acta 1999; 386 281-286.

56. Gomez-Ariza JL, Giraldez I, Sanchez-Rhodes D, Morales E. Metal readsorption and re-distribution during analytical fractionation of trace elements in toxic estuarine sediments. Analytica Chimica Acta 1999; 399 295-307.

57. Cheam V, Lechner J, Sekerka I, Desrosiers R, Nriagu J. Development of laser- excited atomic fluorescence spectrometer and a method for the direct determination of lead in Great Lake waters. Analytica Chimica Acta 1992; 269 129-136.

58. Fischer E, Van D, Berg CMG. Anodic Stripping Voltammetry of Pb and Cd using a Hg film electrode and thiocyanate. Analytica Chimica Acta 1999; 385 273-280.

59. Morales MM, Mart P, Llopis A, Compos L, Sagrado S. An environmental study by factor Analysis of surface sea waters in the Gulf of Valencia (Western Mediterranean). Analytica Chimica Acta 1999; 394 109-117.

60. Hirade M, Chen Z, Sugimoto K, Kawaguchi H. Co precipitation with tin (IV) hydroxide followed by removal of tin carrier for the Determination of trace heavy metals by graphite-furnace atomic absorption Spectrometry. Analytica Chimica Acta 1980; 302 103-107.

61. Ridout PS, Jones HR, Williams JG. Determination of trace elements in a marine reference material of lobster hepatopancreas (TORT-1) using inductively coupled plasma mass spectrometry. Analyst 1988; 113 1383-1386.

62. Sakao SY, OgawaY, Uchida H. Determination of trace elements in seaweed samples by inductively coupled plasma mass spectrometry. Analytica Chimica Acta 1999; 355 121-127.

63. Baeyens W, Monteny F, Leermakers M, Bouillon S. Evalution of sequential extractions on dry and wet sediments. Analytical and Bioanalytical Chemistry 2003; 376 890- 901.

64. Guevara-Riba A, Sahuquillo A, Rubio R, Rauret G. Assessment of metal mobility in dredged harbour sediments from Barcelona, Spain. Science of the Total Environment 2004; 321 241-255.

65. Idris AM, Eltayeb MAH, Potgieter-Vermaak SS, Grieken R, Potgieter JH. Assessment of heavy metal pollution in Sudanese harbours along the Red Sea coast. Microchemical. Journal 2007; 87 104-112.

66. Wepener V, Vermeulen LA. A note on the concentrations and bioavailability of selected metals in sediments of Richards Bay Harbour, South Africa. Water SA 2005; 31 589-595.

67. Esslemont G. Heavy metals in seawater, marine sediments and corals from the Townsville section, Great Barrier Reef Marine Park, Queensland. Marine Chemistry 2000; 71 215- 231.

68. Coetzee PP. Determination and speciation of heavy metals in sediments of the Hartebeespoort Dam By sequential extraction. Water SA 1993; 19 291-300.

69. Salmons W, Förstner U. Trace metal analysis on polluted sediments. Part II:evaluation of Environmental impact. Environmental science and Technology letters 1980; 1 14-24.

70. Li X, Thornton I. Chemical partitioning of trace and major elements in soils contaminated by mining and smelting activities. Applied geochemistry 2000; 16 1693-1706.

71. Kiratli N, Ergin M. Partitioning of heavy metals in surface Black Sea sediments. Applied Geochemistry 1996; 11 775-788.

72. Rauret G, Lopez-Sanchez JF. New sediment and soil CRMs for extractable Trace metal content. International Journal of Environmental Analytical Chemistry 2001; 79 81-95.

73. Salmons W. Adoption of common schemes for single and sequential extractions of Trace metals in soil and sediments. International Journal of Environmental Analytical Chemistry 1993; 51 3-4.

74. Fiedler HD, Lopez-Sanchez JF, Rubio R, Rauret G, Quevauviller PH. Study of the stability of extractable trace metal contents in a river sediment using Sequential extraction. Analyst 1994; 119 1109-1114.

75. Ramos L, Hernandez LM, Gonzalez MJ. Sequential fraction of copper, lead, copper, Cadmium and zinc in soils from or near Donana National Park. Journal of Environmental Quality 1994; 23 7-50.

76. Tu Q, Shan XZ, Ni Z. Evaluation of a sequential extraction procedure for the Fractionationation of amorphous iron and manganese oxides and organic matter in soils. The Science of The total Environment 1994; 151 159-165.

77. Wang S, Jia Y, Wang S, Wang X, Wang H. Fractionation of heavy metals in shallow marine sediments from Jinzhou Bay, China. Journal of Environmental Science (China) 2010 22 23-31.

78. Yuan CG, Shi JB, He B, Liu JF, Liang LN. Speciation of heavy metals in marine sediments from the East China Sea by ICP-MS with sequential extraction. Environment International 2004; 30 769-783.

79. Jones B, Turki A. Distribution and Speciation of heavy metals in surficial sediments from the Tees Estuary, North – East England. Marine Pollution Bulletin 1997; 34 768-779.

80. Ajay SO, Van Loon GW. Studies on redistribution during the analytical fractionation of metals in sediments. The Science of the Total Environment 1989; 87 171-187.

81. Pempkowiak J, Sikora A, Biernacka E. Speciation of heavy metals in marine sediments vs their bioaccumulation by mussels. Chemosphere 1999; 39 313- 321.

82. Forstner U, Wittmann GTW. Metal Pollution in the Aquatic Environment, Springer- Verlag, Berlin. Springer-Verlag, Heidelberg; 1981.

83. Fedotov PS, Zavarzina, a AG, Spivakov BYa, Wennrich, b R, Mattusch J, De K, Titzeb PC, Demin VV. Accelerated fractionation of heavy metals in contaminated soils and sediments using rotating coiled columns, Journal of Environmental Monitoring 2002; 4 318-324.

84. Nemati K, Kartini N, Bakar A, Abas MR, Sobhanzadeh E. Speciation of heavy metals by modified BCR sequential extraction procedure in different depths of sediments from Sungai Buloh, Selangor, Malaysia, Journal of Hazardous Materials 2011; 192(1) 402-410.

85. Mossop KF, Davidson CM. Comparison of original and modified BCR sequential extraction procedures for the fractionation of copper, iron, lead, manganese and zinc in soils and sediments. Analytica Chimica Acta 2003; 478 (1) 111-118.

86. Walkey A, Black TA. An examination of the Dugtijaraff method for determining soil organic matter and proposed modification of the chronic and titration method. Soil Science 1934; 37 23-38.

87. Farnham IM, Johannesson KH, Singh AK, Hodge VF, Stetzenbach KJ. Factor analytical approaches for evaluating groundwater trace element chemistry data. Analytica Chimica Acta 2003; 490 123-138.

88. Ragno G, De Luca M, Ioele G. An application of cluster analysis and multivariate cassification methods to spring water monitoring data. Microchemical Journal 2007; 87 119-127.

89. Jonathan MP, Ram Mohan V. Heavy metals in sediments of the inner shelf off the Gulf of Mannar, Southeast coast of India. Marine Pollution Bulletin 2003; 46 263-268.

90. Sunil Kumar R. Distribution of organic carbon in the sediments of Cochin mangroves, south west coast of India. Indian Journal of Marine Science 1996; 25 274-276.

91. Janaki-Raman D, Jonathan MP, Srinivasalu S, Armstrong-Altrin J S, Mohan SP, Ram-Mohan V. Trace metal enrichments in core sediments in Muthupet mangroves, SE coast of India: Application of acid leachable technique. Environmental Pollution 2007; 145 245-257.

92. Sarkar SK, Bilinski SF, Bhattacharya A, Saha M, Bilinski H. Levels of elements in the surficial estuarine sediments of the Hugli river, northeast India and their environmental implications. Environment International 2004; 30 1089-1098.

93. Tam NFY, Wrong YS. Spatial variation of heavy metalsin surfe sediments of Hong, Kong mangrove swamps. Environmental Pollution 2002; 110 195-205.

94. Subramanian V, Mohanachandran G. Heavy metals distribution and enrichment in the sediments of southern east coast of India. Marine Pollution Bulletin 1990; 21 324-330.

95. Chatterjee M, Massolo S, Sarkar SK, Bhattacharya AK, Bhattacharya BD, Satpathy KK, Saha S. An assessment of trace element contamination in intertidal sediment cores of Sunderban mangrove wetland, India for evaluating sediment quality guidelines. Environmental Monitoring and Assessment 2009; 150 307-322.

96. Datta DK, Subramanian V. Distribution and fractionation of heavy metals in the surface sediments of the Ganges–Brahmaputra–Meghna river system in the Bengal Basin. Environmental Geology 1998; 36 93-101.

97. Müller G. Schwermetalle in den sedimenten des Rheins-Veranderungen seit. Umschan Verlag 1979; 79 133-149.

98. Salomon W, Förstner U. Metals in the hydrocycle. Berlin: Springer; 1984.

99. Buccolieri A, Buccolieri G, Cardellicchio N, Dell'atti A, Leo AD, Maci A. Heavy metals in marine sediments of Taranto Gulf (Ionian Sea, Southern Italy). Marine Chemistry 2006; 99 227-235.

100. Dowling CB, Poreda RJ, Basu AR, Aggarwal PK. Geochemical study of arsenic release mechanisms in the Bengal Basin groundwater. Water Resources Research 2002; 38(9) 1173-1190.

101. Acharyya SK, Lahiri S, Raymahashay BC, Bhowmilk A. Arsenic toxicity of groundwater in parts of the Bengal basin in India and Bangladesh: the role of Quaternary stratigraphy and Holocene sea-level fluctuation. Environmental Geology 2000; 39 231-238.

102. Takarina ND, Browne DR, Risk MJ. Speciation of heavy metals in coastal sediments of Semarang, Indonesia. Marine Pollution Bulletin 2004; 49 854-874.

103. Pickering WF. Metal ion speciation—soil and sediments (a review). Ore Geology Reviews 1986; 1 83-146.

104. Ngiam LS, Lim PE. Speciation patterns of heavy metals in tropical estuarine anoxic and oxidized sediments by different sequential extraction schemes. Science of the Total Environment, 2001; 275 53-61.

105. Dassenakis M, Adrianos H, Depiazi G, Konstantas A, Karabela M, Sakellari A, Scoullos M. The use of various methods for the study of metal pollution in marine sediments, the case of Euvoikos Gulf, Greece. Applied Geochemistry 2003; 18 781-794.

106. Morillo J, Usero J, Gracia I. Heavy metal distri-bution in marine sediments from the southwest coast of Spain. Chemosphere 2004; 55 431-442.

107. Thomas RP, Ure AM, Davidson CM, Littlejoh D, Rauret G, Rubio R, López-Sánchez JF. Three-stage sequential extraction procedure for the deter-mination of metals in river sediments. Analytica Chimica Acta 1994; 286 423-429.

108. Panda D, Subramanian V, Panigrahy RC. Geochemical fractionation of heavy metals in Chilka Lake (east coast of India) – a tropical coastal lagoon. Environmental Geology 1995; 26 199-210.

109. Dawson EJ, Macklin MG. Speciation of heavy metals in floodplain and flood sediments: a reconnaissance survey of the Aire Valley, West Yorkshire, Great Britain Environmental Geochemistry and Health 1998; 20 67-76.

110. Ramos L, González M, Hernández L. Sequential extraction of copper, lead, cadmium, and zinc in sediments from Ebro River (Spain): relationship with levels detected in earthworms. Bulletin of Environmental Contamination and Toxicology 1999; 62 301-308.

111. Caille N, Tiffreau C, Leyval C, Morel JL. Solubility of metals in an anoxic sediment during prolonged aeration. Science of the Total Environment 2003; 301 239-250.

112. Canfield DE. Reactive iron in marine sediments. Geochimica et Cosmochimica Acta 1989; 53 619-632.

113. Petersen W, Wallman K, Li PL, Schroeder F, Knauth HD. Exchange of trace elements at the sediment– water interface during early diagenesis processes. Marine and Freshwater Research 1995; 46 19-26.

114. Monbet P. Mass balance of lead through a small macrotidal estuary: the Morlaix River estuary (Brittany, France). Marine Chemistry 2006; 98 59-80.

115. Adriano DC. Trace elements in terrestrial environments. New York: Springer; 1986.

116. Sarkar SK, Saha M, Takada H, Bhattacharya A, Mishra P, Bhattacharya B. Water quality management in the lower stretch of the river Ganges, east coast of India: an approach through environmental education. Journal of Cleaner Production 2007; 15 1559-1567.

117. Banerjee ADK. Heavy metal levels and solid phase speciation in street dusts of Delhi, India. Environmental Pollution 2003; 123(1) 95-105.

118. Long ER, Ingersoll CG, MacDonald DD. Calculation and uses of mean sediment quality guideline quotients: a critical review. Environmental Science and Technology 2006; 40 1726-1736.

119. Venkatramanan S, Ramkumar T, Anithamary I, Jonathan MP. Speciation of selected heavy metals geochemistry in surface sediments from Tirumalairajan river estuary, east coast of India. Environmental Monitoring and Assessment 2013; 185(8) 6563-6578.

120. Ranjan RK, Singh G, Routh J, Ramanathan AL. Trace metal fractionation in the Pichavaram mangrove–estuarine sediments in southeast India after the tsunami of 2004. Environmental Monitoring and Assessment 2013 (article in press).

121. Mohan M, Augustine T, Jayasooryan KK, Chandran MSS, Ramasamy EV. Fractionation of selected metals in the sediments of Cochin estuary and Periyar River, southwest coast of India, Environmentalist 2012; 32 383-393.

122. Hseu ZY. Extractability and bioavailability of zinc over time in three tropical soils incubated with biosolids. Chemosphere 2006; 63 762-771.

123. Perin G, Craboledda L, Lucchese M, Cirillo R, Dotta L, Zanette ML, Orio AA. Heavy metal speciation in the sediments of Northern Adriatic Sea- a new approach for environmental toxicity determination, in: T.D. Lekkas (Ed.), Heavy Metal in the Environment 1985; 2 454-456.

124. Dhanakumar S, Murthy KR, Solaraj G, Mohanraj R. Heavy-Metal Fractionation in Surface Sediments of the Cauvery River Estuarine Region, Southeastern Coast of India, Arch Environmental Contamination Toxicology 2013; 65 14-23.

125. Dezileau L, Pizarro C Rubio MA. Sequential extraction f iron in marine sediments from the Chilen continental margin. Marine Geology 2007; 241 111-116.

Chapter 6

PHOSPHORUS ADSORPTION OF SOME BRAZILIAN SOILS IN RELATIONS TO SELECTED SOIL PROPERTIES

Valdinar Ferreira Melo[1], Sandra Cátia Pereira Uchôa[1], Zachary N. Senwo[2], and Ronilson José Pedroso Amorim[3]

[1]Department of Soil and Agricultural Engineering, Federal University of Roraima, Boa Vista, Brazil

[2]Department of Biological & Environmental Sciences, Alabama A&M University, Huntsville, USA

[3]Agronomy, Federal University of Roraima, Boa Vista, Brazil

ABSTRACT

A major nutritional problem to crops grown in highly weathered Brazilian soils is phosphorus (P) deficiencies linked to their low availability and the capacity of the soils to fix P in insoluble forms. Our studies examined factors that might influence P behavior in soils of the Amazon region. This study was conducted to evaluate the maximum phosphate adsorption capacity (MPAC) of the soils developed from mafic rocks (diabase), their parent materials and other factors resulting in the formation of eutrophic soils having A chernozemic horizon associated with Red Nitosols (Alfisol) and Red Latosols (Oxisol) of the Amazonian environment. The MPAC was determined in triplicates as a function of the remnant P values. The different concentrations used to determine the MPAC allowed maximum adsorption values to be reached for all soils. The Latosol (Oxisol) and Nitosol (Alfisol) soils presented higher phosphate adsorption values that were attributed to the oxidic mineralogy and high clay texture while the Chernosol (Mollisol) soils presented the lowest pho- sphate adsorption values.

INTRODUCTION

Due to the climatic variations impacting the Amazon region of Brazil, the Northeast area of Roraima has transitioned from a semi-arid climatic condition to one with high precipitation [1]. The increase in precipitation has greatly affected the soil's biogeochemical weathering attributes, erosion and deposition processes, thus contributing to the formation of deep, highly weathered soils dominant in kaolinite and oxidic mineralogy [2]. Some areas of the Amazon have soils originated from mafic rocks (diabase), resulting in the formation of Eurtophic Red Nitosol (Haplustalf), Dystrophic Yellow Red Latosol (Xanthic Haplustox), Eutropic Haplic Tb Cambisols, with A chernozem, Orthic black Chernosol (Mollisol) and vertic Orthic Ebanic Cheronosol (Mollisol) soil series [3].

Studies have shown that the soils consist typically of kaolinitic mineralogy and low fertility, with the predominant classes being Yellow Latosols (Oxisol) and Yellow Argisols (Ultisol) [4] -[6]. The indigenous areas of Roraima in general are located where these soils are found and reflect certain aspects of the sustainability and quality of life of the indigenes cultivating the soils for agricultural production. In the Raposa Serra do Sol Indigenous Reserve (approximately 1,747,464 ha); the proportion of soils (approximately 19,900 ha) considered as fertile for sustainable agricultural production is situated on gentle undulating relief and includes diabase soils of the "Pedra Preta" Sill [7].

A major nutritional problem to crops grown in these soils is P deficiency linked to low available P content and capacity to fix fertilizer P in insoluble form. Evaluating the P availability in these soils is important because soil P adsorption capacity directly influences plant response to phosphate fertilization of soils. Phosphorus fixation in soils, widely designated as adsorption, is a slow process that can take years to reach a balance and may lead to decrease in P availability. In most Brazilian soils, studies have shown that the major factors influencing soil P adsorption are the clay fractions, mineralogy, amorphous colloid content, pH, exchangeable aluminum and organic matter [8] -[10].

Phosphorus adsorption depends on the nature and quality of sites available on the mineral surfaces and is therefore affected by high clay contents within the same mineralogy [11]. Studies by Ker et al. [9] indicate that the mineralogy, crystal size, and specific surfaces of soils will be more important than the quantity of clay in determining soil's P adsorption capacities [12] -[14]. Bedin et al. [14] state that the presence of large proportions of sesquioxides on clays influences phosphate adsorption and precipitation with iron and aluminum. The objective of this study was to evaluate the maximum P adsorption capacity (MPAC) of the soils developed from mafic rocks (diabase) in northeastern Roraima and its relationships with certain physical, chemical and mineralogical attributes.

MATERIALS AND METHODS

The soils used in this study were sampled from a toposequence and consisted of five soil profiles, developed from mafic rocks that form one of the largest mafic rock bodies in North Amazon's "Pedra Preta" Sill. The Maloca do Flechal study areas (Figure 1) is inhabited by the Macuxi Indians and situated in the northeast of the state of Roraima, Brazil. The climate average about 1200 mm in rainfall, with sunshine duration of 12 hours/day. The region's relief is characterized as undulated with step savannah vegetation and patches of seasonal forest on the East-West transect line.

The fieldwork consisted of opening trenches in a toposequence, collecting the soil samples in each profile soil and describing them according to Santos et al. [15]. Soils were classified based on the Brazilian Soil Classification System manual published by Embrapa [16] and analyzed for chemical properties (Table 1) as described in a soil analysis manual published by Embrapa [17]. The profiles were classified as Eutrophic Red Nitosol (Alfisol)- NVe, Tb Eutrophic Haplic Cambisol (Inceptisol)-CXbe; Orthic Ebanic Chernosol (Mollisol)-MEo; vertic Orthic Ebanic Chernosol (Mollisol)-MEov and Distgrophic Yellow Red Latosol (Oxisol)-LVAdf.

The Fe associated with the low crystal minerals in the surface horizons was extracted with ammonium oxalate [18], and the free Fe extracted using dithionite-citrate bicarbonate (DCB) [19]. The Fe content was determined using induced plasma emission spectrometry while the Fe associated with the crystalline minerals was calculated by the difference between the Fe extracted by DCB (Fe_d) and the Fe extracted with oxalate (Fe_o). The mineralogy was determined by X-ray diffraction of the clay and silt fraction. From the natural, Fe-free clay fraction, oriented slides were prepared and irradiated in an X-ray diffractometer at sweeping angle (2θ) intervals between $2°$ and $40°$ and a goniometer speed of $2°$ in 2θ/min, using $CuK\alpha$ radiation with Ni filter [20]. Powdered silt slides were analyzed under the same conditions while the diffraction patterns were interpreted according to Chen [21].

Phosphorus (P) was evaluated as described by Alvarez and Fonseca [22] using 5 g of fine ground soils mixed with a solution of $CaCl_2$ (0.01 mol·L^{-1}) containing 50 mg·L^{-1} P and shaken for an hour then filtered and the P concentration using Murphy and Riley [23] procedure.

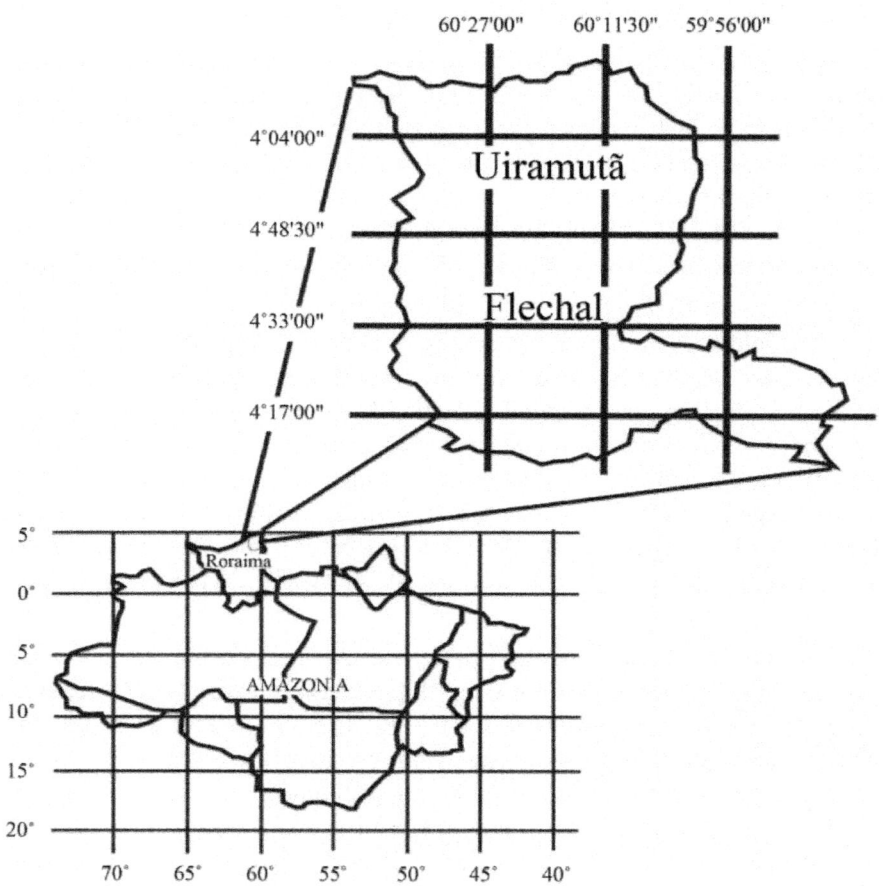

Figure 1. The maloca do flechal study areas.

Table 1. Chemical characteristics of soils studied.

Soil	Hor.	pH	Ca²⁺	Mg²⁺	Na⁺	K⁺	Al³⁺	H+Al	S	CEC	V	SOM	P
			---------------------cmol$_c$·kg⁻¹---------------------					(%)				g·kg⁻¹	mg·kg⁻¹
NVe	Ap	5.7	2.63	2.48	0.13	0.10	0.05	6.87	5.34	12.21	43	24.6	0.96
	Bnit	5.9	1.03	0.91	0.03	0.01	-	2.20	1.98	4.18	65	9.3	0.72
MEo	Ap	6.1	6.87	3.16	0.32	0.15	-	4.34	10.50	14.84	71	23.8	2.50
	Bnit	6.2	5.80	2.62	0.04	0.05	0.05	3.57	8.51	12.08	71	14.2	1.01
CXbe	Ap	6.8	8.69	2.79	0.11	0.05	0.05	2.75	11.64	14.39	81	15.3	1.33
	Bi₂	7.3	9.21	0.00	0.12	0.05	-	1.54	9.38	10.92	86	9.5	1.44
MEov	Ap	7.0	9.31	0.04	0.09	0.05	-	2.53	9.35	11.88	79	20.6	1.16
	Bi₂	7.4	12.7	0.06	0.11	0.05	-	1.43	12.92	14.35	90	3.7	1.39
CXbe	Ap	6.5	6.75	0.37	0.07	0.04	-	3.24	7.23	10.63	69	19.3	1.22
	Bi₂	7.8	7.31	0.00	0.09	0.03	-	0.71	7.43	8.14	91	3.7	1.08
LVAd	A	5.4	0.00	0.49	0.14	0.11	0.50	6.60	0.74	7.34	10	24.6	1.48
	Bw	4,3	0.00	0.50	0.04	0.04	0.50	8.52	0.52	9.04	06	22.5	0.53

Hor. = Horizon. 1) pH in water. 2) S = sum of bases. CEC = Cation exchange capacity at pH 7.0. V = percentage of saturation by bases; SOM = Soil organic matter. P = Phosphorus extractable by Mehlich-1. NVe-Red Nitosol. MEo-Orthic Ebanic Chernosol. Cxbe-Tb Eutrophic Haplic Cambisol. MEov-Vertic Orthic Ebanic Chernosol, LVAd-Distrophic Red Yellow Latosol.

The Langmuir isotherm were obtained by adding 25 mL $CaCl_2$ (10 mol·L^{-1}) solution containing P in doses ranging from 0 to 60 mg·L^{-1}. After 24 hours agitation, the samples were centrifuged and the P in the equilibrium solution was quantified colorimetrically [23].

The data were fitted to the non-linear Langmuir model, relating the concentration of the adsorbed element per adsorption unit (soil) and the concentration of the element in the equilibrium solution (supernatant), and the maximum adsorption capacity calculated. The sorption values of each soil were plotted according to the Langmuir isotherm: $C/(x/m) = (1/Kb) + (C/b)$. Where C, is equilibrium P concentration (mg·L^{-1}), x/m is the amount of sorbed P (mg·kg^{-1}), m and b, are the constants related to P sorption maximum (mg·kg^{-1}) and K is the bonding energy (L·mg^{-1}), respectively. The plot of C/(x/m) versus C should give a straight line from which 1/b (slope) and K (slope or intercept) can be calculated [24]. The Pearson analysis of simple linear correlation using the Statistic program 7.0 between MPAC and some soil characteristics was carried out, and the correlations greater than 95% were considered significant ($\alpha < 0.05$).

RESULTS AND DISCUSSION

The Fe contents determined from successive extractions using DCB (2), ranged from 29.4 (MEov) to 258.7 g·kg^{-1} (LVAdf). The values seemed high and atypical of Roraima [5] [6] and Amazonia soils [25] [26], given that the predominant soils in these environments have iron contents averaging less than 73 g·kg^{-1}. The LVAdf stood out among the soils studied with lower Fe_o/Fe_d ratio indicating the predominance of high crystalline iron oxide.

The soils mineralogical compositions (Table 2) showed the presence of high mineral activities. Illite was common in all the soil horizons but smectite was restricted to soils with the A Chernozem horizon (CXbe and MEov). Kaolinite also occurred generally in most of the profiles, indicating intense weathering with strong lixiviation of bases, low fertility and lower (CEC) especially for LVAdf. The mineralogy of the 2:1 group might have resulted from the geology and climatic variations in the Amazon over geological time. The northeast of Roraima has passed through a recent sub-period of semi-arid climate to the current conditions of greater precipitations [27] [28]. However, this mineralogy contrasts with the data obtained by Vale Júnior [5] and Melo et al. [6] for other Roraima soils.

The P values varied greatly among the soil profiles and horizons. The highest values were observed in the surface horizons (Table 3), ranging from 19.8 (LVAdf) to 50.2 mg·L^{-1} (MEo). For the surface horizons, the values ranged from 6.6 mg·L^{-1} (LVAdf) to 46.9 (MEov) and this could be attributed to the type of clay mineralogy present in the soils. According to Alvarez et al. [29], the soils and horizons analyzed presented low P retention, except for LVAdf, which could be attributed to the highly negative charge on the smectites surfaces of the clays. The high P values in most of the soils could be explained by the striking presence of 2:1 minerals. Alves and Lavorenti [30] worked with Latosols in the state of São Paulo, Brazil and reported the lack of correla-

Table 2. Mineralogical composition of the clay, silt and sand fractions of soils studied.

Soil	Hor.	Clay	Silt	Sand
NVe	Ap	Ct, Gt	Qz, Ct, Bt, Fs, Ilm	Qz, Mi
	Bnit	Ct, Il, Gt	Qz, Ct, Bt, Fs, Ilm, Pg	Qz, Mi,
MEo	Ap	Ct, Il, Gt	Qz, Bt, Fs, Px	Qz, Mi
	Bnit	Ct, Il, Gt	Qz, Ct, Mi	Qz, Mi
CXbe	Ap	Ct, Il/Mt, Es, Gt	Qz, Fs, Ct, Px, Af	Qz, Fs, Pg, Mi
	Bi$_2$	Ct, Il, Es	Pg, Fs, Qs, Ct, Mi, Af	Pg, Qz, Fs, Mi, Mg, Cn
MEov	Ap	Ct, Il, Es	Fs, Pg, Qz, Ct, Mi, Af	Qz, Pg, Af, Fs
	Bi$_2$	Es, Ct, Il	Ac, Cl, Mi, Fs	Qz, Pg, Fs, Cn
CXbe	Ap	Ct, Il, Gt	Ct, Qz, Zl, Pg	Qz, Fs, Ru
	Bi$_2$	Ct, Il, Gt	Ct, Qz, Hb, Bt	Qz
LVAd	Ap	Gb, Ct, Il, Gt	Qz, Ct, Gb, Ru, Hm	Qz
	Bw$_1$	Ct, Gb, Il	Qz, Ct, Gb	Qz

Qz-Quartz, Ct-Kaolinite, Es-Smectite, Fs-Feldspar, Mi-Mica, Pg-Plagioclase, Af-Amphiboles, Px-Piroxene, Il-Illite, Bt-Biotite, Ac-Actinolite, Cl-Chlorite, Zl-Zeolite, Hb-Hornblendite, Ru-Rutile, Gb-Gibbsite, Gt-Goethite; Hm-Hematite, Mg-Manganite, Cn-Corundum.

Table 3. Maximum phosphate adsorption capacity for the soils studied.

Soil	Horizon	Clay g·kg^{-1}	P-remaining mg·L^{-1}	MAPC mg·kg^{-1}
NVe	Ap	520	39.8	439.0
	Bnit.	640	4.2	1280.0
MEo	Ap	280	50.2	353.0
	Bnit.	370	36.2	475.0
CXbe	Ap	320	43.3	409.0
	Bi$_2$	470	29.5	551.0
MEov	Ap	290	47.9	370.0
	Bi$_2$	290	46.9	378.0
CXbe	Ap	350	40.7	431.0
	Bi$_2$	380	37.7	460.0
LVAdf	Ap	650	19.8	701.0
	Bw$_1$	700	6.6	1111.0

Nve-Red nitosol, MEo-Orthic ebanic chernosol, Cxbe-Tb eutrophic haplic cambisol, MEov-vertic orthic ebanic chernosol LVAdf-distrophic red latosol.

tions between P and goethite and P and kaolinite. High iron-sesquioxides contents are often considered as the determining factor of P adsorption in soils [31]; however, the low crystalline forms should fix more P than the crystalline forms [30]. The differences in behavior observed for LVAdf was more related to the presence of gibbsite than to the presence of the low crystalline forms, considering the low Fe_o/Fe_d ratio presented by this soil (Table 4).

The MPAC values (Table 3) ranged from 353 to 701 mg·kg^{-1} in the surface horizons and from 378 to 1280 mg·kg^{-1} for the sub-surface horizons. Except for the values obtained in the MEov profile and the MEo surface layer, the rest of the soils were in the range of MPAC values as indicated by Novais and Smyth [32] for Cerrado Latosols. In Amazon soils, Lima et al. [26] reported MPAC values ranging from 210 to 2170 mg·kg^{-1} in Oxisol, while Singh et al. [33] obtained values between 160 and 980 mg·kg^{-1} P. For central Amazon, Falcão and Silva [34] reported MPAC ranging from 298 to 888 mg·kg^{-1} for Ultisols. The MPAC values were lower than those normally found in highly weathered soils. In basalt-derived soils, Oxisol and Alfisol, Bognola [35] found MPAC values between 1620 and 2740 mg·kg^{-1}, while Valladares et al. [36] obtained values between 526 and 1.667 mg·kg^{-1} for Ultisols. Soils originating from this material have higher MPAC because of the higher clay content and occurrence of oxides such as hematite and goethite [37] [38]. Despite the parental materials, climate and relief contributed to the soil formation with high smectite contents in undulated relief, highly weathered rich Fe and Al oxides contents. The structure, chemical compositions, exchangeable ion type and small crystal size of smectite clays are responsible for the unique properties, including a large chemically active surface area, a high cation exchange capacity and low anion exchange capacity.

In the high clay soils (Table 3), phosphate adsorption was more pronounced. The Yellow Red Latosol (Oxisol) presented the greatest MPAC in the two horizons, which can be explained by the high clay content, high Fe_d and low Fe_c contents (Table 4). The Nitosol also presented high MPAC, greater than LVAdf, in the subsurface horizon, which can be due to the high clay content. Vertic Orthic Ebanic Chernosol had the least adsorbed P in the two horizons, which can be explained by the low clay contents and low Fe_d and Fe_c values. Generally the soils showed increase in P adsorption with increase in soil depth, which can be attributed to increase in the clay contents from horizon A to horizon B. The high phosphate adsorption in the Latosol and Nitosol in the Amazon soils can be attributed to the low Fe contents predominate in this environment [6].

The clay, Fe_d and Fe_c contents correlated positively and significantly with the MPAC, suggesting that these attributes contributed to an increase in the MPAC in the horizons and profiles studied (Table 5). Given the strict correlations between MPAC and clay, Fe_d and Fe_c contents, it can be postulated that the presence of goethite in all the soils was a major determinant for the MPAC and not kaolinite soils. Bahia Filho et al. [39] tested the effects of the mineralogical components on MPAC and observed that goethite (Gt) was mainly responsible, contributing 6% of the total MPAC.

Table 4. Soil iron and silicon dioxide contents of soils.

Soil	Hor.	Fe_d	Fe_o	Fe_o/Fe_d	Fe_c	SiO_2
		$g·kg^{-1}$				
NVe	Ap	58.3	3.5	0.06	54.8	1.41
	Bnit	75.5	3.2	0.04	72.3	1.13
MEo	Ap	106.0	7.7	0.07	98.3	1.26
	Bnit	95.1	5.6	0.06	86.9	1.90
CXbe	Ap	52.0	6.4	0.12	45.6	2.01
	Bi_2	51.1	4.9	0.10	46.2	2.65
MEov	Ap	34.9	7.7	0.22	27.2	2.01
	Bi_2	29.4	1.6	0.05	27.8	1.37

CXbe	Ap	77.1	7.2	0.09	69.9	1.61
	Bi_2	64.8	2.5	0.04	62.3	2.03
LVAdf	A	180.3	1.8	0.01	178.9	0.73
	Bw	258.7	1.0	0.004	257.7	0.88

Nve-Red nitosol, MEo-Orthic ebanic chernosol, Cxbe-Tb eutrophic haplic cambisol, MEov-vertico orthic ebanic chernosol, LVAdf-distrophic red yellow latosol. Fe_d-Dithionite-extracted Fe. Fe_o-Oxalate-extracted iron. Fe_c-crystalline Fe.

Table 5. Simple correlations between maximum phosphate adsorption capacity of soils and some soils properties.

	A Horizon	B Horizon
Clay	0.79*	0.94**
T	−0.87*	−0.79*
C	0.38[n.s.]	0.54[n.s.]
Fe_d	0.83*	0.93**
Fe_o	−0.87*	−0.62*
Fe_c	0.84*	0.94**

T-Clay. Fe_d-Dithionite-extracted Fe. Fe_o-Oxalate-extracted iron. Fe_c-crystalline Fe. n.s.-not significant. *mean of $0.01 <$ probability < 0.05; **significant at probability < 0.01.

The high affinity of geothite soils for P is likely due to the easy access of surface phosphate anions. The $H_2PO_4^-$ occupies site of the hydroxides (OH^-) previously coordinated to Fe^{3+}, forming much more stable surface complexes.

The correlation between the soil organic C and MPAC was not significant for the soils studied. This might be because organic C can influence P adsorption positively in numerous ways [40], negatively [41] or no effect [42]. The latter has been observed in Cerrado ecosystem (Savanna) soils where the C stock increases linearly with the soil clay and silt contents, demonstrating that total C does not increase the P adsorption, but enhances the richness of the minerals that retain P.

The CEC values and Fe_o correlated significantly and negatively, respectively, with MPAC. Indeed P adsorption tends to diminish with increase in CEC in the soils, because the negative charges repel the phosphate ion. On the other hand, negative correlation of MPAC with Fe_o is in line with reported studies. Hernández and Meurer [43] studied three forms of soils in Uruguay, an Argiudoll, a Hapludert, and a Natraqualf, and observed positive correlations between P adsorption and low Fe crystalline forms (Fe_o).

CONCLUSION

The Oxisol and the Alfisol soils presented the highest phosphate adsorption values, due to Fe and Al oxides rich minerals. The Molissol soils showed the lowest adsorption values due to the high presence of smectite clay. The soils that showed positive correlations with MPAC were in the order: A Horizon-Fe_c > Fe_d > clay and B Horizon-clay = Fe_c > Fe_d. The soils of the northern border of the Amazon exhibit P adsorption behavior different from other areas of the Amazon. Sustaining agricultural production in these soils will require management approaches that will enhance P availability. An adequate pool of labile P in these soils could be enhanced with inorganic P fertilizations and other organic residues.

REFERENCES

1. Schaefer, C.E.R. and Dalrymple, J. (1996) Pedogenesis and Relict Properties of Soils with Columnar Structure from Roraima, North Amazonia. Geoderma, 71, 1-17.http://dx.doi.org/10.1016/0016-7061(95)00073-9

2. Benedetti, U.G., Vale Jr., J.F., Schaefer, C.E.G.R., Melo, V.F. and Uchôa, S.C.P. (2011) Genesis, Chemistry and Mineralogy of Soils Derived from Plio-Plestocene Sediments and from Volcanic Rocks in Roraima—North Amazonia. Brazilian Journal of Soil Science, 35, 299-312.

3. Melo, V.F., Schaefer, C.E.G.R. and Uchôa, S.C.P. (2010) Indian Land Use in the Raposa-Serra do Sol Reserve, Roraima, Amazonia, Brazil: Physical and Chemical Attributes of a Soil Catena Developed from Mafic Rocks under Shifting Cultivation. Catena, 80, 95-105.http://dx.doi.org/10.1016/j.catena.2009.09.004

4. Schaefer, C.E.G.R., Resende, S.B., Correia, G.F. and Lani, J.L. (1993) Chemical and Minerological Characteristics of Sodium-Afected Soils from Northeeastern Roraima. Brazilian Journal of Soil Science, 17, 431-438.

5. Vale Jr., J.F. (2000) Pedogenesis and Changing of Soil under Shifting Cultivation in Areas of Acid and Basic Volcanic Rocks in Northeastern Roraima. Ph.D. Thesis, Universidade Federal de Viçoa, Viçosa.

6. Melo, V.F., Schaefer, C.E.G.R., Fontes, L.E.F., Chagas, A.C., Lemos Jr., J.B. and Andrade, R.P. (2006) Physical Chemical and Mineralogical

Characteristics of Soils from the Agricultural Colony of Apiaú (Roraima, Amazonia), under Different Land Uses and after Burning. Brazilian Journal of Soil Science, 30, 1039-1050.

7.   Melo, V.F. (2002) Soil and Indicators of Agricultural Use in Roraima: Indigenous Areas of Maloca Flechal and Colonization of Apiaú. Ph.D. Thesis, Universidade Federal de Viçoa, Viçosa.

8.   Brennan, R.F., Bolland, M.D.A., Jeffery, R.C. and Allen, D.G. (1994) Phosphorus Adsorption by a Range of Western Australian Soils Related to Soil Properties. Communications in Soil Science and Plant Analysis, 25, 2785-2795.http://dx.doi.org/10.1080/00103629409369225

9.   Motta, P.E.F., Curi, N., Siqueira, J.O., Van Raij, B., Furtini Neto, A.E. and Lima, J.M. (2002) Adsorption and Forms of Phosphorus in Latosols: Influence of Mineralogy and Use. Brazilian Journal of Soil Science, 26, 349-359.

10.   Moreira, F.L.M., Mota, F.O.B., Clemente, C.A., Azevedo, B.M. and Bomfim, G.V. (2006) Phosphorus Adsorption in Ceará State Soils, Brazil. Revista Ciência Agronômica, 37, 7-12.

11.   Bahia Filho, A.F.C., Braga, J.M., Resende, M. and Ribeiro, A.C. (1983) Relationship between Phosphorus Adsorption and Mineralogical Components of Latosols Clay Fraction of the Central Plateau. Brazilian Journal of Soil Science, 7, 221-226.

12.   Torrent, J., Barron, V. and Schwertmann, U. (1990) Phosphate Adsorption and Desorption by Goethites Differing in Crystal Morphology. Soil Science Society of American Journal, 54, 1007-1012. http://dx.doi.org/10.2136/sssaj1990.03615995005400040012x

13.   Ruan, H.D. and Gilkes, R.J. (1996) Kinetics of Phosphate Sorption and Desorption by Synthetic Aluminous Goethite before and after Thermal Transformation to Hematite. Clay Minerals, 31, 63-74. http://dx.doi.org/10.1180/claymin.1996.031.1.06

14.   Bedin, I., Furtini Neto, A.E., Resende, A.V., Faquin, V., Tokura, A.M. and Santos, J.Z.L. (2003) Phosphate Fertilizers and Soybean Production in Soils with Different Phosphate Buffer Capacities. Brazilian Journal of Soil Science, 27, 639-646.

15.   Santos, R.D., Lemos, R.C., Santos, H.G., Ker, J.C. and Anjos, L.H. (2005) Description Handbook and Collecting of Soil in Field. 5 Edition, Sociedade Brasileira de Ciência do Solo, Viçosa, MG.

16.   Embrapa (Empresa Brasileira de Pesquisa Agropecuaria) (2006) Brazilian System of Soil Classification. 2nd Edition, Embrapa Solos, Rio de Janeiro.

17. Embrapa (Empresa Brasileira de Pesquisa Agropecuaria) (1997) Manual of Soil Analysis Methods. 2nd Edition, National Soil Research Center, Rio de Janeiro.

18. Mckeague, J.A. and Day, J.H. (1966) Dithionite and Oxalate Extractable Fe and Al as Aids in Differentiating Various Classes of Soils. Canadian Journal of Soil Science, 46, 13-22.http://dx.doi.org/10.4141/cjss66-003

19. Mehra, O.P. and Jackson, M.L. (1960) Iron Oxide Removal from Soils and Clay by a Dithionite-Citrate System Buffered with Sodium Bicarbonate. Clays and Clay Minerals, 7, 317-327. http://dx.doi.org/10.1346/CCMN.1958.0070122

20. Whittig, L.D. and Allardice, W.R. (1986) X-Ray Diffraction Techniques. In: Klute, A., Ed., Methods of Soil Analysis, Part 1: Physical and Mineralogical Methods, American Society of Agronomy, Madison, 331-362.

21. Chen, P.-Y. (1977) Table of Key by Lines in X-Ray Power Diffraction Patterns of Minerals in Clays and Associated Rocks. Occasional Paper 21, Department of Natural Resources, Geological Survey, Bloomington, 67 p.

22. Alvarez, V.V.H. and Fonseca, D.M. (1990) Definition of Doses of Phosphorus to Determine the Maximum Adsorption Capacity for Phosphate in Greenhouse Trials. Brazilian Journal of Soil Science, 14, 49-55.

23. Murphy, J. and Riley, J.P. (1962) A Modified Single Solution Method for the Determination of Phosphate in Natural Waters. Analytica Chimica Acta, 27, 31-36.http://dx.doi.org/10.1016/S0003-2670(00)88444-5

24. Bera, R., Seal, A., Bhattacharyya, P., Mukhopadhyay, K. and Giri, R. (2006) Phosphate Sorption Desorption Characteristics of Some Ferruginous Soils of Tropical Region in Eastern India. Environmental Geology, 51, 399-407. http://dx.doi.org/10.1007/s00254-006-0335-9

25. Silva, J.R.T. (1999) Soils of Acre: Physical, Chemical and Mineralogical and Phosphate Adsorption. Thesis of Ph.D., Federal Universtiy of Viçosa, MG, Brazil.

26. Lima, H.N., Melo, J.W.V., Schaefer, C.E.G.R., Ker, J.C. and Lima, A.M.N. (2006) Mineralogy and Chemistry of Three Soils along a Toposequence from the Upper Solimões Basin, Western Amazonia. Brazilian Journal of Soil Science, 30, 59-68.

27. Schaefer, C.E.G.R. (1994) Soils and Paleosols from Northeast Roraima, North Amazonia: Geomorphology, Genesis and Landscape Evolution. Ph.D. Dissertation, University of Reading, Reading, Berkshire.

28. Schaefer, C.E.G.R. and Vale Júnior, J.F. (1997) Climate Changing and Landscape Evolution in Roraima: A Review from Cretaceous to Recent. In: Barbosa, R.I., Ferreira, E.J.G. and Castellón, E.G., Eds., Man, Environment and Ecology in the State of Roraima, INPA, Manaus, 231-293.

29. Alvarez, V.V.H., Novais, R.F., Dias, L.E. and Oliveira, J.A. (2000) Determination and Use of the Remaining Phosphorus. Newsletter, Sociedade Brasileira de Ciência do Solo, 25, 27-33.

30. Alves, M.E. and Lavorenti, A. (2004) Remaining Phosphorus and Sodium Fluoride pH in Soils with Different Clay Contents and Clay Mineralogies. Pesquisa Agropecuária Brasileira, 39, 241-246. http://dx.doi.org/10.1590/S0100-204X2004000300006

31. Uehara, G. and Gillman, G. (1981) The Mineralogy, Chemistry and Physics of Tropical Soils with Variable Charge Clays. Westview, Boulder.

32. Novais, R.F. and Smyth, T.J. (1999) Phosphorus in Soil and Plant in Tropical Conditions. Universidade Federal de Viçosa-MG, Viçosa, 399 p.

33. Singh, R., Moller, M.R.F. and Ferreira, W.A. (1983) Phosphorus Adsorption Kinetics in Amazonian Soils. Revista Brasileira de Ciência do Solo, 7, 227-231.

34. Falcão, N.P. and Silva, J.R.A. (2003) Phosphorus Adsorption Characteristics in Some Central Amazonian Soils. Acta Amazônica, 34, 337-342.http://dx.doi.org/10.1590/S0044-59672004000300001

35. Bognola, I.A. (1995) Chemical, Physical and Mineralogical Soil Intermediaries between Latosols Brunos and Latosols Purple. Master's Thesis, Universidade Federal de Viçosa, Viçosa.

36. Valladares, G.S., Pereira, M.G. and Anjos, L.H.C. (2003) Phosphate Sorption in Low Activity Clay Soils. Bragantia, 62, 111-118. http://dx.doi.org/10.1590/S0006-87052003000100014

37. Curi, N., Camargo, O.A., Guedes, G.A.A. and Silveira, J.V. (1988) Phosphorus Sorption in Latosols of Southeast and South of Brazil. In: Reunião de Classificação e Correlação de Solos e Interpretação de Aptidão Agrícola, 3rd Edition, Empresa Brasileira de Pesquisa Agropecuária, Rio de Janeiro, 150.

38. Fontes, M.P.F. and Weed, S.B. (1996) Phosphate Adsorption by Clays from Brazilian Oxisols: Relationships Whit Specific Surface Area and Mineralogy. Geoderma, 72, 37-51.http://dx.doi.org/10.1016/0016-7061(96)00010-9

39. Bahia Filho, A.F.C., Vasconcellos, C.A., Santos, H.L., Mendes, J.F., Pitta, G.V.E. and Oliveira, A.C. (1982) Inorganic Phosphorus and Available

Phosphorus Forms in a Dark Red Latosol, Fertilized with Different Phosphates. Brazilian Journal of Soil Science, 6, 99-104.

40. Parfitt, R.L. (1978) Anion Adsorption by Soils and Soil Materials. Advances in Agronomy, 30, 1-50. http://dx.doi.org/10.1016/S0065-2113(08)60702-6

41. Borggaard, O.K., Jorgensen, S.S., Moberg, J.P. and Raben-Lage, B. (1990) Influence of Organic Matter on Phosphate Adsorption by Aluminum and Iron Oxides in Sandy Soils. European Journal of Soil Science, 41, 443-449. http://dx.doi.org/10.1111/j.1365-2389.1990.tb00078.x

42. Zinn, Y.L., Lal, R. and Resck, D.V.S. (2005) Texture and Organic Carbon Relations Described by a Profile Pedotransfer Function for Brazilian Cerrado Soils. Geoderma, 127, 168-173. http://dx.doi.org/10.1016/j.geoderma.2005.02.010

43. Hernández, J. and Meurer, E.J. (1990) Phosphorus Sorption in Soils from Uruguay and Its Relationship with Iron Oxides. Brazilian Journal of Soil Science, 22, 223-230.

Chapter 7

FLEXURAL BEHAVIOR OF LATERALLY LOADED TAPERED PILES IN COHESIVE SOILS

Musab Aied Qissab

Department of Civil Engineering, Al-Nahrain University, Baghdad, Iraq

ABSTRACT

In this paper, the flexural behavior of laterally loaded tapered piles in cohesive soils is investigated. The exact solution for the governing differential equation of the problem is obtained based on the beam-on-elastic foundation approach in which the soil reaction on the pile is related directly to the pile lateral deflection. In this investigation, the modulus of subgrade reactions is assumed to be constant along the pile depth. Parametric study through numerical examples is carried out to prove the validity and accuracy of the obtained results. In general, the derived displacement field can be used to study pile response in multilayered soil profiles by subdividing the pile into a number of elements. It is found that tapered piles show stiffer behavior than that for prismatic ones having the same material volume with an optimum stress distribution along the pile depth. Accordingly, tapered piles are more efficient and economic than those having the same material volume. Verification is also carried out for the obtained results through finite element analysis and the selected number of elements gives a very good agreement for lateral deflection and a larger number of elements is required to obtain better results for bending moment because of moment loss resulting from the lack of shear diagram.

INTRODUCTION

Piles are widely used to support structures not limited to bridges, high rise buildings and offshore structures which are subjected toaxial and lateral loads resulting from different sources. Tapered piles, as special cases, have received great attention at present due to their good performance in resisting loads in comparison to that for prismatic ones because of the optimum material distribution with respect to loading intensity. Most of the available analysis and design guidelines lay more emphasis on prismatic piles over tapered piles despite of the economical advantage of the latter. Tapered piles are not widely used as a design option because of the limit knowledge about their behavior under different loading types in comparison to the prismatic piles.

There are a number of studies concerning the behavior of individual piles. Wei [1] studied experimentally the static behavior of piles in cohesionless soils under the effect of axial, lateral, and cyclic loads. Two sets of tests with three types of geometries including a prismatic pile in dry sandy soil were conducted to study their behavior. The results of the study confirmed the efficiency of tapered piles over the prismatic ones having the same material input.

Horvath et al. [2] investigated experimentally the behavior of tapered tube piles under axial, uplift, and lateral loads in sand. The experimental program was mainly conducted for one of the larger transportation projects for the major renovation and expansion of John F. Kennedy International Airport in New York City to verify the performance of these piles. It was established from the experimental results that taper-tube piles are successfully resist the entire spectrum of axial and lateral loads that is normally encountered in transportation engineering.

Shankar et al. [3] developed a procedure to predict the flexural behavior of axially loaded and laterally loaded tapered piles embedded in liquefaction-induced laterally spread soils. The problem was analyzed by using the modulus of subgrade reaction approach based on Winkler type soil model. The resulting governing equation to solve the flexural behavior of the pile with the specified boundary conditions was solved by using finite difference technique. The use of tapered piles was found beneficial in liquefaction-induced laterally spreading soils as the maximum bending moment developed due to drag force is less especially when the applied axial force is much lower than the critical load.

Zhan et al. [4] studied the load capacity behavior of two series of axially loaded tapered piles in sand by using finite element method. It was observed from the numerical analysis that the shaft resistance increasing with the tapered angle with an increase of (12%) over that of the straight-side piles at an optimum tapered angle. It was concluded that tapered piles are more suitable for floating pile foundations.

STATEMENT OF THE PROBLEM

The tapered pyramidal pile shown in Figure 1 of length (L) and embedded in a homogeneous cohesive soil layer. is subjected at its head to a lateral concentrated load (Q) and a bending moment (M).

The governing differential equation of the above problem was given by Hetenyi [5] for beams on elastic foundation with variable flexural rigidity as follows:

$$\frac{d^2}{dz^2}\left[EI_{(z)}\frac{d^2y}{dz^2}\right]+k_z y=0$$

(1)

Or

$$E\left[I_{(z)}\frac{d^4y}{dz^4}+2\left(\frac{dI_{(z)}}{dz}\right)\frac{d^3y}{dz^3}+\left(\frac{d^2I_{(z)}}{dz^2}\right)\frac{d^2y}{dz^2}\right]+k_z y=0$$

(2)

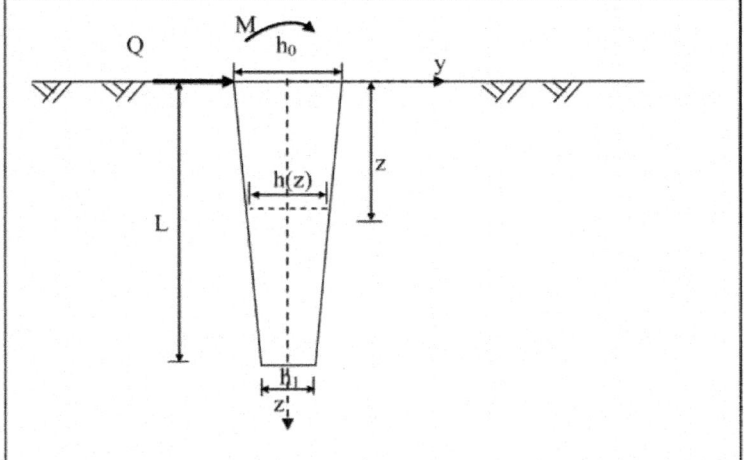

Figure 1: The considered pile configuration.

where k_z is the modulus of subgrade reaction $= Kb(z)$, and E is the modulus of elasticity of the pile material. The moment of inertia I(z) can be given by:

$$I_{(z)}=\frac{b_{(z)}\left(h_{(z)}\right)^3}{12}=\frac{b_0 h_0^3}{12}\left(1-cz\right)^4=I_0\left(1-cz\right)^4$$

(3)

in which:

$b_{(z)}=b_0\left(1-cz\right)$, $h_{(z)}=h_0\left(1-cz\right)$; b_0, h_0 are the section dimensions of the pile at z = 0; $c=(h_0-h_1)/(h_0 L)$;

h_1 = the pile dimension in the direction of y at z = L; and I_0 is the moment of inertia at z = 0.

Substituting Equation (3) into (2) and assuming that $\overline{z} = (1-cz)$ and after simplifications, the following equation can be obtained:

$$c^4\overline{z}^4\frac{d^4y}{d\overline{z}^4} + 8c^4\overline{z}^3\frac{d^3y}{d\overline{z}^3} + 12c^4\overline{z}^2\frac{d^2y}{d\overline{z}^2} + \frac{k_z}{EI_0}y = 0$$

(4)

The solution of Equation (4) can be expressed as follows:

$$y(z) = A_1\overline{z}^{(a_1+b_1(i))} + A_2\overline{z}^{(a_2+b_2(i))} + A_3\overline{z}^{(a_3+b_3(i))} + A_4\overline{z}^{(a_4+b_4(i))}$$

(5)

or

$$y(z) = A_1\overline{z}^{(a_1)}e^{i(b_1 \ln(\overline{z}))} + A_2\overline{z}^{(a_2)}e^{i(b_2 \ln(\overline{z}))} + A_3\overline{z}^{(a_3)}e^{i(b_3 \ln(\overline{z}))} + A_4\overline{z}^{(a_4)}e^{i(b_4 \ln(\overline{z}))}$$

(6)

where

$$\left.\begin{aligned}
a_1 &= \frac{1}{2}(-1-a), b_1 = -\frac{\overline{\beta}}{a}\\
a_2 &= \frac{1}{2}(-1+a), b_2 = \frac{\overline{\beta}}{a}\\
a_3 &= \frac{1}{2}(-1-a), b_3 = \frac{\overline{\beta}}{a}\\
a_4 &= \frac{1}{2}(-1+a), b_4 = -\frac{\overline{\beta}}{a}
\end{aligned}\right\}$$

(7)

$$a = \sqrt{\frac{5}{2}\left[1+\sqrt{1+\left(\frac{4}{5}\overline{\beta}\right)^2}\right]}, \ \overline{\beta} = \sqrt{\frac{\beta^4}{c^4}-1}, \ \beta = \frac{k_z}{EI_0}, \ \text{and} \ \overline{z} = (1-cz)$$

(8)

The constants A_1, A_2, A_3, and A_4 can be determined by applying the boundary conditions as follows:

at $(z = 0)$

$$\frac{d^2y}{dz^2} = \frac{M}{EI_0}$$

(9)

$$E\left[I_{(z)}\frac{d^3y}{dz^3} + \frac{dI_{(z)}}{dz}\frac{d^2y}{dz^2}\right] = Q$$

(10)

or

$$\left[(\bar{z})\frac{d^3y}{dz^3}-4c\frac{d^2y}{dz^2}\right]=\frac{Q}{EI_0}$$

(11)

at $(z = L)$

$$\frac{d^2y}{dz^2}=0$$

(12)

$$\left[(\bar{z})\frac{d^3y}{dz^3}-4c\frac{d^2y}{dz^2}\right]=0$$

(13)

Applying the above boundary conditions, the following equations can be obtained:

at $(z = 0)$

$$\left(\bar{z}^{a_1}D_{21}-2a_1cD_{11}+a_1(a_1-1)c^2D_{01}\right)A_1+\left(\bar{z}^{a_1}D_{22}-2a_1cD_{12}+a_1(a_1-1)c^2D_{02}\right)A_2$$
$$+\left(\bar{z}^{a_2}D_{21}-2a_2cD_{11}+a_2(a_2-1)c^2\right)A_3+\left(\bar{z}^{a_2}D_{21}-2a_2cD_{11}+a_2(a_2-1)c^2\right)A_4=M/(EI_0)$$

(14)

$$(\bar{z}D_{311}-4cD_{211})A_1+(\bar{z}D_{312}-4cD_{212})A_2+(\bar{z}D_{321}-4cD_{221})A_3+(\bar{z}D_{322}-4cD_{222})A_4=\frac{Q}{EI_0}$$

(15)

at $(z = L)$

$$\left(\bar{z}^{a_1}D_{21}-2a_1c\bar{z}^{(a_1-1)}D_{11}+a_1(a_1-1)c^2\bar{z}^{(a_1-2)}D_{01}\right)A_1$$
$$+\left(\bar{z}^{a_1}D_{22}-2a_1c\bar{z}^{(a_1-1)}D_{12}+a_1(a_1-1)\bar{z}^{(a_1-2)}c^2D_{02}\right)A_2$$
$$+\left(\bar{z}^{a_2}D_{21}-2a_2c\bar{z}^{(a_2-1)}D_{11}+a_2(a_2-1)c^2\bar{z}^{(a_2-2)}D_{01}\right)A_3$$
$$+\left(\bar{z}^{a_2}D_{22}-2a_2c\bar{z}^{(a_2-1)}D_{12}+a_2(a_2-1)c^2\bar{z}^{(a_2-2)}D_{02}\right)A_4=0$$

(16)

$$(\bar{z}D_{311}-4cD_{211})A_1+(\bar{z}D_{312}-4cD_{212})A_2+(\bar{z}D_{321}-4cD_{221})A_3+(\bar{z}D_{322}-4cD_{222})A_4=0$$

(17)

where:

$$D_{01} = \sin\left(\frac{\bar{\beta}}{a}\ln(\bar{z})\right)$$

$$D_{02} = \cos\left(\frac{\bar{\beta}}{a}\ln(\bar{z})\right)$$

$$D_{11} = -\frac{\bar{\beta}}{a}\left(\frac{c}{\bar{z}}\right)\cos\left(\frac{\bar{\beta}}{a}\ln(\bar{z})\right)$$

$$D_{12} = \frac{\bar{\beta}}{a}\left(\frac{c}{\bar{z}}\right)\sin\left(\frac{\bar{\beta}}{a}\ln(\bar{z})\right)$$

$$D_{21} = -\gamma^2 \sin\left(\frac{\bar{\beta}}{a}\ln(\bar{z})\right) - \frac{\bar{\beta}}{a}\left(\frac{c}{\bar{z}}\right)^2 \cos\left(\frac{\bar{\beta}}{a}\ln(\bar{z})\right)$$

$$D_{22} = -\gamma^2 \cos\left(\frac{\bar{\beta}}{a}\ln(\bar{z})\right) + \frac{\bar{\beta}}{a}\left(\frac{c}{\bar{z}}\right)^2 \sin\left(\frac{\bar{\beta}}{a}\ln(\bar{z})\right)$$

$$\tag{18}$$

and

$$D_{211} = \bar{z}^{a_1} D_{21} - 2a_1 c\bar{z}^{(a_1-1)} D_{11} + a_1(a_1-1)c^2\bar{z}^{(a_1-2)} D_{01}$$

$$D_{212} = \bar{z}^{a_1} D_{22} - 2a_1 c\bar{z}^{(a_1-1)} D_{12} + a_1(a_1-1)c^2\bar{z}^{(a_1-2)} D_{02}$$

$$D_{221} = \bar{z}^{a_2} D_{21} - 2a_2 c\bar{z}^{(a_2-1)} D_{11} + a_2(a_2-1)c^2\bar{z}^{(a_2-2)} D_{01}$$

$$D_{222} = \bar{z}^{a_2} D_{22} - 2a_2 c\bar{z}^{(a_2-1)} D_{12} + a_2(a_2-1)c^2\bar{z}^{(a_2-2)} D_{02}$$

$$D_{311} = \bar{z}^{a_1} D_{31} - 3a_1 c\bar{z}^{(a_1-1)} D_{21} + 3a_1(a_1-1)c^2\bar{z}^{(a_1-2)} D_{11} - a_1(a_1-1)(a_1-2)c^3\bar{z}^{(a_1-3)} D_{01}$$

$$D_{312} = \bar{z}^{a_1} D_{32} - 3a_1 c\bar{z}^{(a_1-1)} D_{22} + 3a_1(a_1-1)c^2\bar{z}^{(a_1-2)} D_{12} - a_1(a_1-1)(a_1-2)c^3\bar{z}^{(a_1-3)} D_{02}$$

$$D_{321} = \bar{z}^{a_2} D_{31} - 3a_2 c\bar{z}^{(a_2-1)} D_{21} + 3a_2(a_2-1)c^2\bar{z}^{(a_2-2)} D_{11} - a_2(a_2-1)(a_2-2)c^3\bar{z}^{(a_2-3)} D_{01}$$

$$D_{322} = \bar{z}^{a_2} D_{32} - 3a_2 c\bar{z}^{(a_2-1)} D_{22} + 3a_2(a_2-1)c^2\bar{z}^{(a_2-2)} D_{12} - a_2(a_2-1)(a_2-2)c^3\bar{z}^{(a_2-3)} D_{02}$$

$$\tag{19}$$

in which,

$$D_{31} = \gamma^3 \cos\left(\frac{\bar{\beta}}{a}\ln(\bar{z})\right) - 2\left(\frac{\bar{\beta}}{a}\right)^2\left(\frac{c}{\bar{z}}\right)^3 \sin\left(\frac{\bar{\beta}}{a}\ln(\bar{z})\right) - \gamma^2\left(\frac{c}{\bar{z}}\right)\sin\left(\frac{\bar{\beta}}{a}\ln(\bar{z})\right) + 2\frac{\bar{\beta}}{a}\left(\frac{c}{\bar{z}}\right)^3 \cos\left(\frac{\bar{\beta}}{a}\ln(\bar{z})\right)$$

$$D_{32} = -\gamma^3 \sin\left(\frac{\bar{\beta}}{a}\ln(\bar{z})\right) - 2\left(\frac{\bar{\beta}}{a}\right)^2\left(\frac{c}{\bar{z}}\right)^3 \cos\left(\frac{\bar{\beta}}{a}\ln(\bar{z})\right) - \gamma^2\left(\frac{c}{\bar{z}}\right)\cos\left(\frac{\bar{\beta}}{a}\ln(\bar{z})\right) + 2\frac{\bar{\beta}}{a}\left(\frac{c}{\bar{z}}\right)^3 \sin\left(\frac{\bar{\beta}}{a}\ln(\bar{z})\right)$$

$$\tag{20}$$

$$\gamma = \left(\frac{\bar{\beta}}{a}\right)\left(\frac{c}{\bar{z}}\right)$$

$$\tag{21}$$

NUMERICAL EXAMPLES

In this paper, the behavior of two groups of piles (with square cross-section) is investigated. The geometry, loading condition and materials constants for each group are presented in Figure 2 andFigure 3.

The lateral deflection, the distribution of shear force and bending moment along the pile shaft are given in Figures 4-6.

The results of group No. 2 are given in Figures 7-9. Pile (S2) in this group has the same material volume for pile (T3).

It can be observed from Figure 4 and Figure 7 that tapered piles show stiffer behavior than prismatic ones having the same material volume. The deflection decrease for a tapering angle (0.955°) (pile T1) is found to be

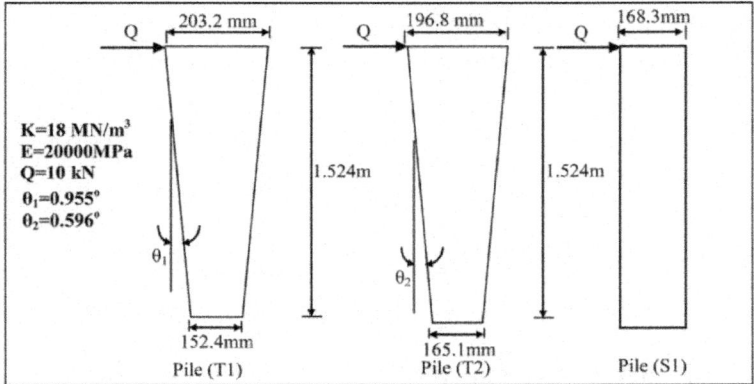

Figure 2: Pile group No. 1.

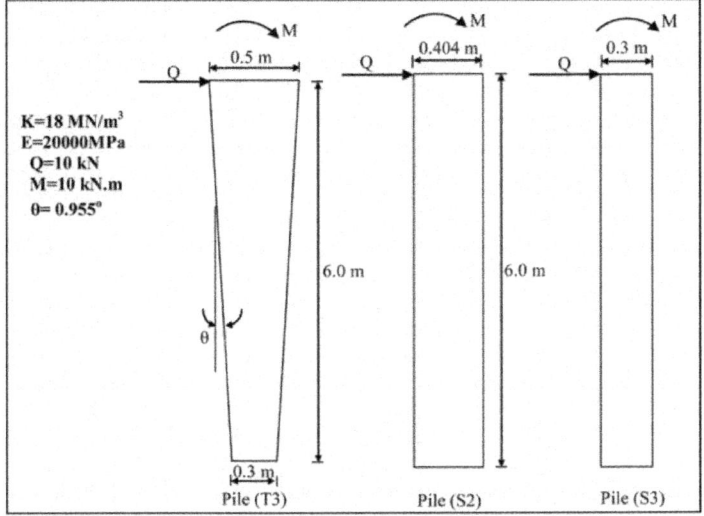

Figure 3: Pile group No. 2.

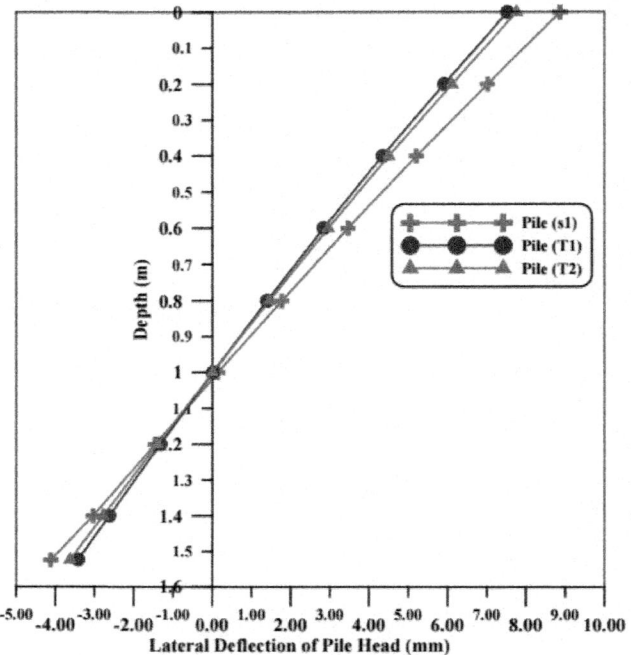

Figure 4: Lateral deflection of pile group No. 1.

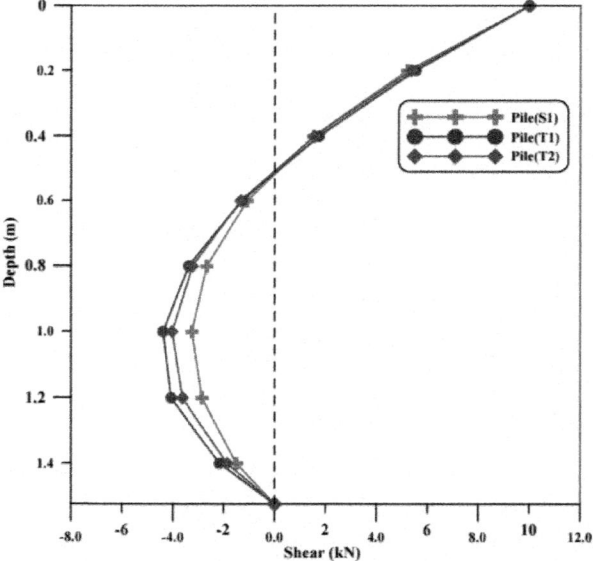

Figure 5: Shear force distribution for pile group No. 1.

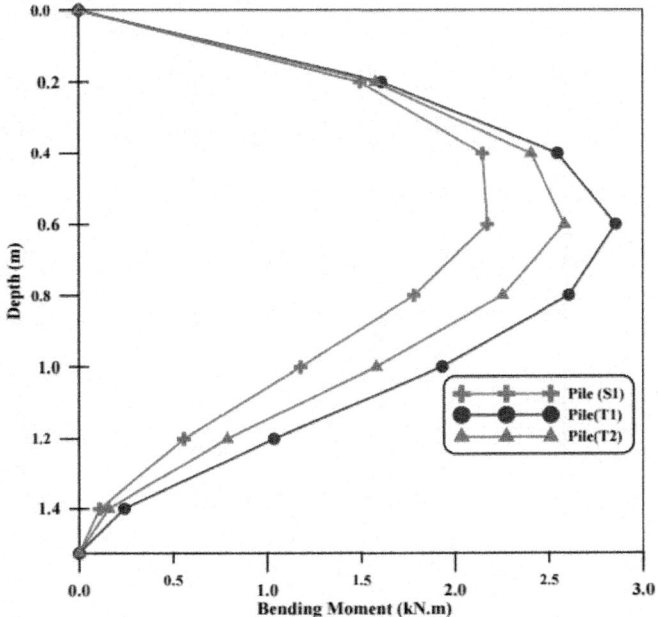

Figure 6: Moment distribution for pile group No. 1.

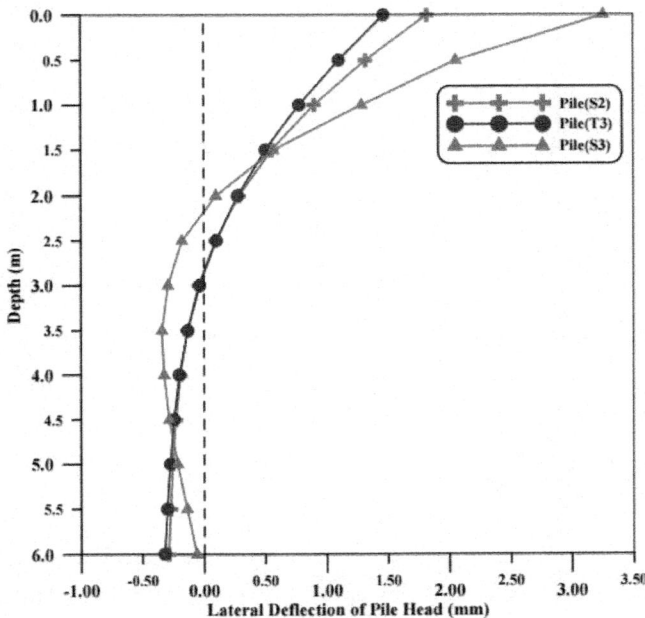

Figure 7: Lateral deflection for pile group No. 2.

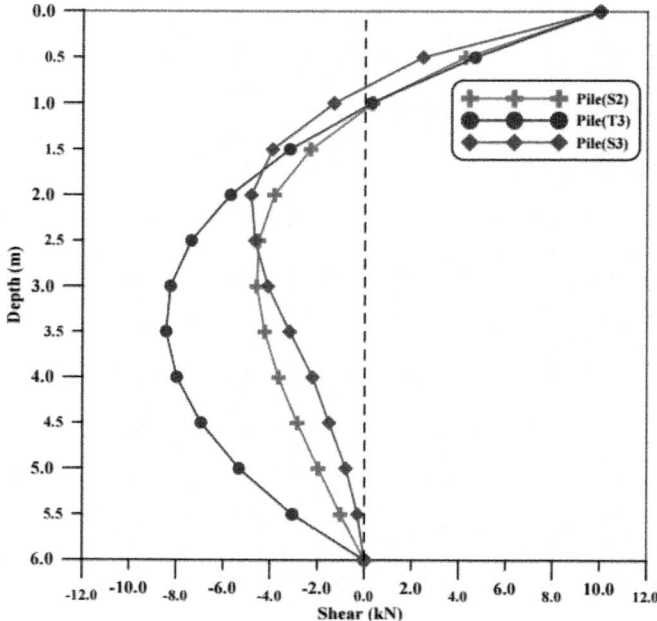

Figure 8: Shear force distribution for pile group No. 3.

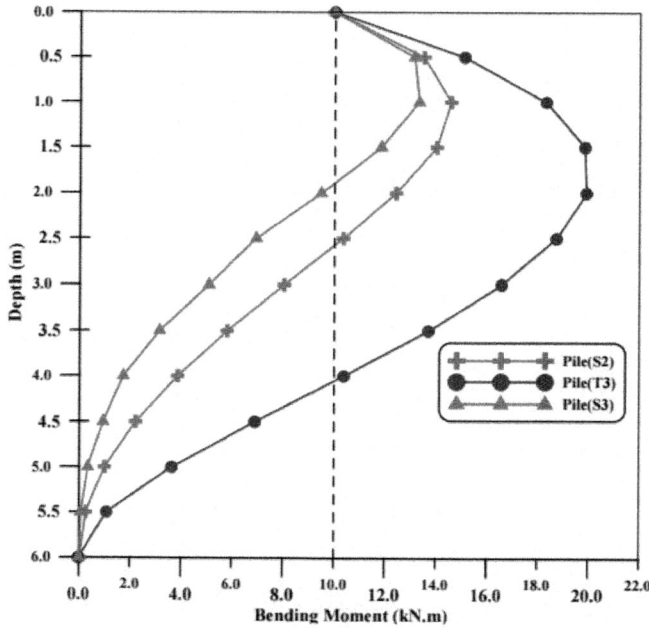

Figure 9: Moment distribution for pile group No. 3.

(17.84%) for group No. 1 and (22%) for group No. 2 (pile T3) for the same material volume. Also, it can be noted that the point of maximum bending moment is located approximately at the upper third of the pile depth at which the cross-sectional area of the tapered piles is larger than that for prismatic ones having the same material volume. This gives a more efficient material distribution.

VERIFICATION OF THE PRESENT SOLUTION

To verify the obtained results from the derived equations and because there is no laboratory studies that concerned with laterally loaded tapered piles in cohesive soils, the finite element method is used to analyze piles T1 and T3 of the two groups. Each pile is subdivided into a number (10) of straight elements as shown in Figure 10 and a computer program is used to solve the problem.

It can be noted from the Figure 11 and Figure 13 that the lateral deflection curves obtained from the two solutions are identical. On the other hand, the moment distribution curves shown in Figure 12 and Figure 14 obtained by the finite element method give a lower-bound solution for the bending moment. This is due to the lack in shear diagram resulting from subdividing the pile into a number of straight elements which cause a loss in bending moment (the area of the shear diagram). To obtain better results for the bending moment, finer divisions for the pile should be used. This verifies the accuracy and efficiency of the proposed solution in comparison to the finite element method.

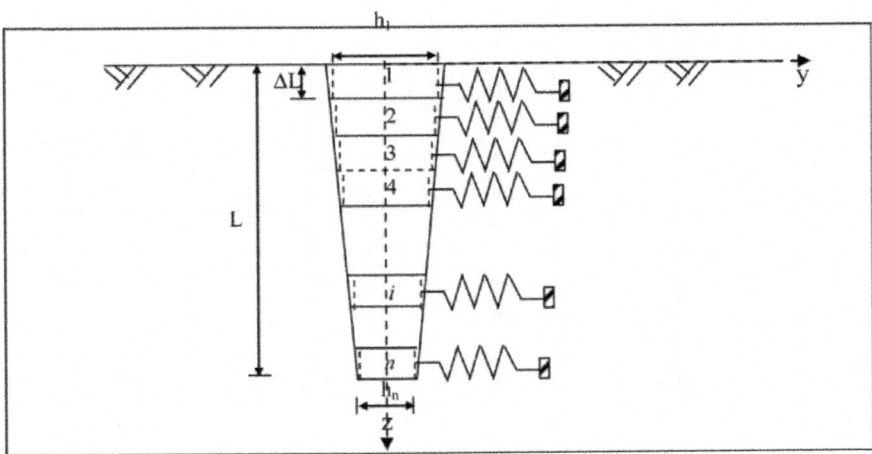

Figure 10: The proposed finite element model of the tapered pile.

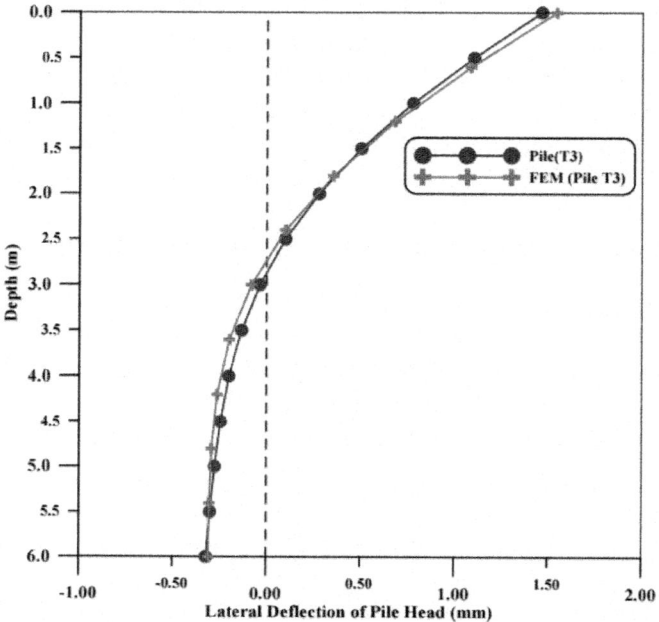

Figure 11: Lateral deflection for piles T4 and S4.

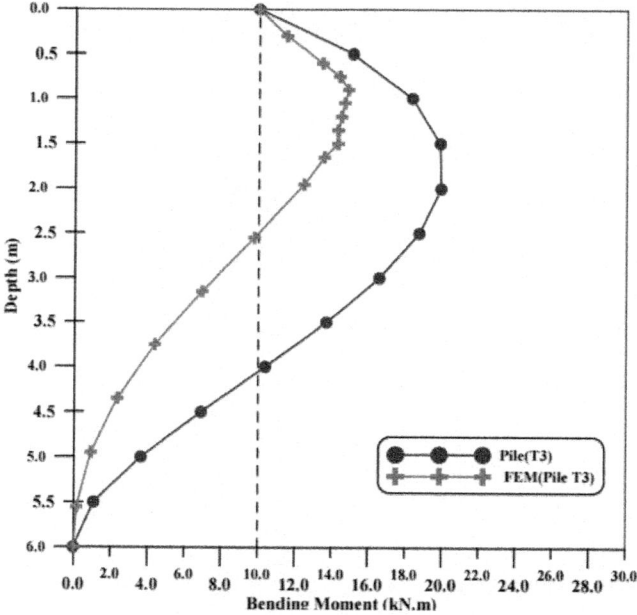

Figure 12: Moment distribution for piles T4 and S4.

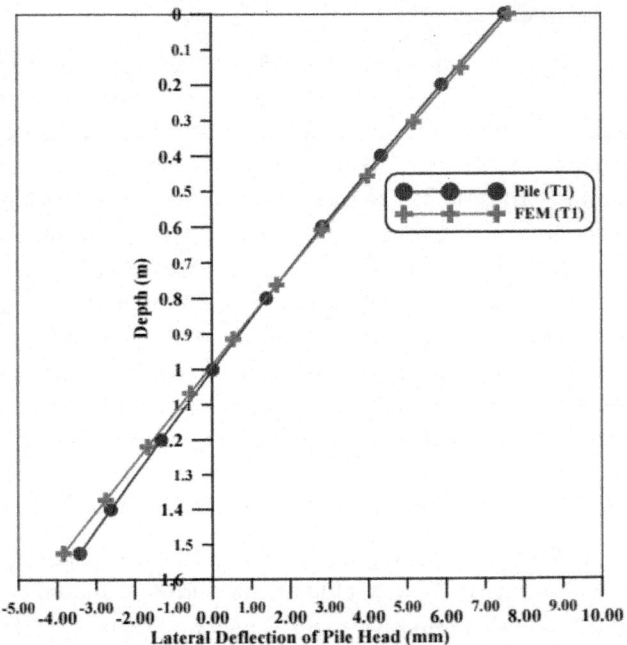

Figure 13: Lateral deflection for pile T1.

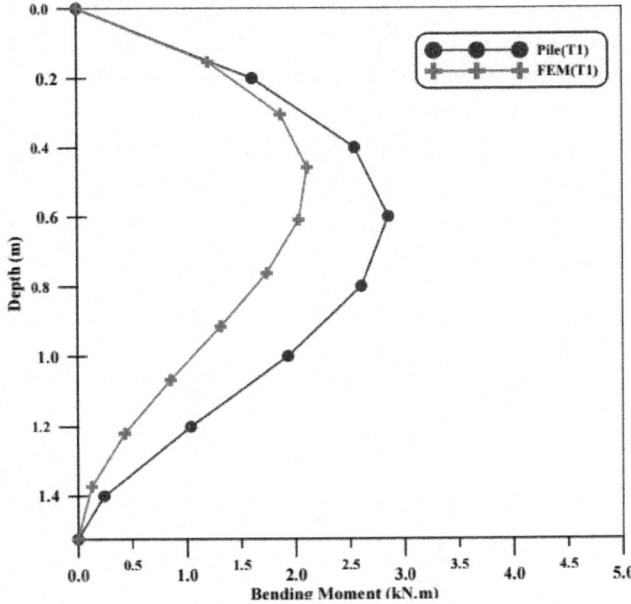

Figure 14: Moment distribution for pile T1.

CONCLUSION

In this paper, an exact solution of the differential equation for a pyramidal tapered pile with square cross-sec- tional area in cohesive soils has been obtained based on beam-on-elastic foundation theory assuming constant value for the subgrade reaction coefficient. It is clear from the presented results that tapered piles show stiffer behavior than that for prismatic ones having the same material volume. It is found that the decrease in deflection for a tapering angle $0.955°$ is in the range of 17.84% - 22% for the studied cases. The value of the maximum bending moment for tapered piles is found to be larger than that for prismatic piles. The increase in maximum bending moment for tapered piles is proportional to the increase in cross-section size which gives approximately constant bending stress. As a result, tapered piles are more efficient and economic than those having the same material volume.

REFERENCES

1. Wei, J.Q. (1998) Experimental Investigation of Tapered Piles. M.Sc. Thesis, University of Western Ontario, London, Canada.

2. Horvath, J.S., Trochalides, T., Burns, A. and Merjan, S. (2004) A New Type of Tapered Steel Pipe Pile for Transportation Applications. Geo-Trans, American Society of Civil Engineers Geo-Institute, California.

3. Shanker, K., Basudhar, P.K. and Patra, N.R. (2008) Flexural Response of Tapered Piles in Liquefied Soils. The 12th International Conference of International Association for Computer Methods and Advances in Geomechanics (IACMAG), Goa.

4. Zhan, Y., Wang, H. and Liu, F. (2012) Numerical Investigation on Load Capacity Behavior of Tapered Pile Foundations. EJGE, 17, 1969-1980.

5. Hetenyi, M. (1946) Beams on Elastic Foundations. University of Michigan Press, Ann Arbor.

Chapter 8

INFLUENCE OF MECHANICAL PROPERTIES OF CONCRETE AND SOIL ON SOLICITATIONS OF MAT FOUNDATION

Oustasse Abdoulaye Sall[1], Makhaly Ba[1], Mapathé Ndiaye[1], Daouda Sangare[2], Mathioro Fall[1], and Alassane Thiam[1]

[1]Département de Génie Civil, UFR SI-Université de Thiès, Thiès, Sénégal

[2]Département de Mathématiques, UFR SAT-Université Gaston Berger, Saint Louis, Sénégal

ABSTRACT

This work studies the influence of mechanical and geometrical characteristics of the concrete and the soil on the stresses in a mat foundation. In this study, the soil-structure interaction is modeled by two parameters, the modulus of subgrade vertical reaction (k) and the modulus of subgrade horizontal reaction (2T). These two parameters are dependent on the geometrical and mechanical characteristics of the system. Results of this study show a sensitivity of solicitations to variations of geometrical and mechanical characteristics of the model. Although solicitations in the plate are sensitive to mechanical properties of concrete, these solicitations are strongly influenced by the mechanical and geometrical characteristics of the soil mass. However, it should be noted that the influence of E_b is denoted in the center of the plate whereas the E_s feels almost in the same manner over the entire extent of the plate. This study also shows that for the same load cases, the values of the torsion moment and shear stress are not significant those of bending moments and normal stresses, respectively.

INTRODUCTION

Developments in the construction of civil engineering and especially disorders observed in the supporting structures push the practitioners to better take into account soil-structure interaction in the process of calculating the foundation structures. In addition to this, the structural and geotechnical calculations of foundations are generally conducted separately, that's why it would be interesting to develop a computational approach that would take into account the geotechnical and structural considerations related to the problematic of the calculation of foundation. It is in this context that an accurate and complete characterization is needed for reliable calculation, hence the interest of this research. A foundation is responsible for transmitting loads from the superstructure to the soil. It provides an interface between the upper part of the structure and the soil. A mat foundation can be considered as a reinforced concrete slab over the whole of the structure whose study is mainly governed by the plate theory [1]. The behavior can be studied from the Lagrange equation with inclusion of the soil-structure interaction. The resolution of the behavioral model (theory of plate and soil-structure interaction) is possible with the use of methods of double Fourier series, numerical analysis (finite differences and finite elements) with boundary conditions defined [1]. This research studies the influence of mechanical properties of concrete and soil foundation on stresses generated by a uniformly distributed load applied on the plate.

MODELING THE RAFT ON SOIL MASS

Modeling of the Concrete Structure

A mat foundation is considered as a reinforced concrete slab resting on the soil mass. Usually, the thickness is small compared to other dimensions. It is in this context that the Kirchhoff model [2] is adopted in this study.

Mat foundations are structures in planar state constraints. They admit vertical displacements along the z axis and the behavioral model may be governed by the model of Kirchhoff for the plates [2]. The plate can be re-presented by the following Figure 1.

Considering the deflection (w) as known, the behavioral model of a plate supported on its periphery can be given by the following equation:

$$\frac{\partial^4 w}{\partial x^4} + 2\frac{\partial^4 w}{\partial x^2 \partial y^2} + \frac{\partial^4 w}{\partial y^4} = q(x,y)/D$$

(1)

where D is the flexural rigidity of the plate and is given by:

$$D = \frac{E_b e^3}{12\left(1-v_b^2\right)}$$

(2)

with:

E_b: elastic modulus of the material constituting the plate;

e: the thickness of the plate;

v_b: Poisson's ratio of the plate.

Modeling of Soil Mass

To model the subgrade, the Filonenko-Borodich biparametric model was selected. The model of Filonenko- Borodich [3] [4] provides continuity between the springs of the Winkler model by a thin elastic membrane under constant tension T (Figure 2) which connects the springs. Settlement (w) of the soil surface under pressure (q) is given by:

$$q(x,y) = k \cdot w(x,y) - T \cdot \nabla^2 w(x,y)$$

(3)

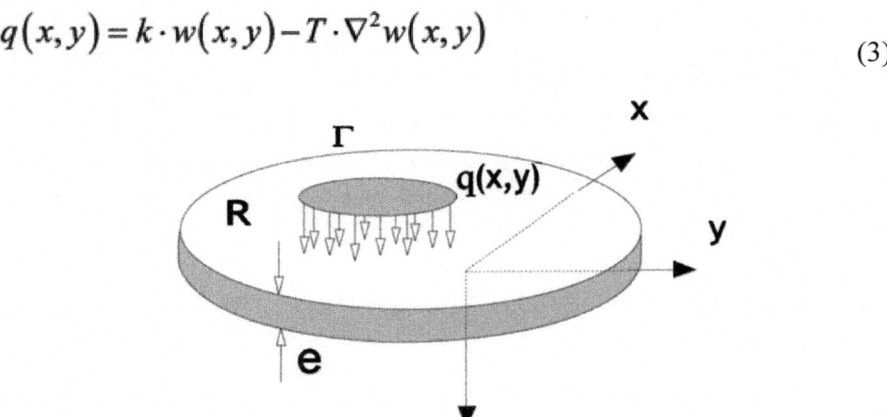

Figure 1. Geometry of the plate and external forces.

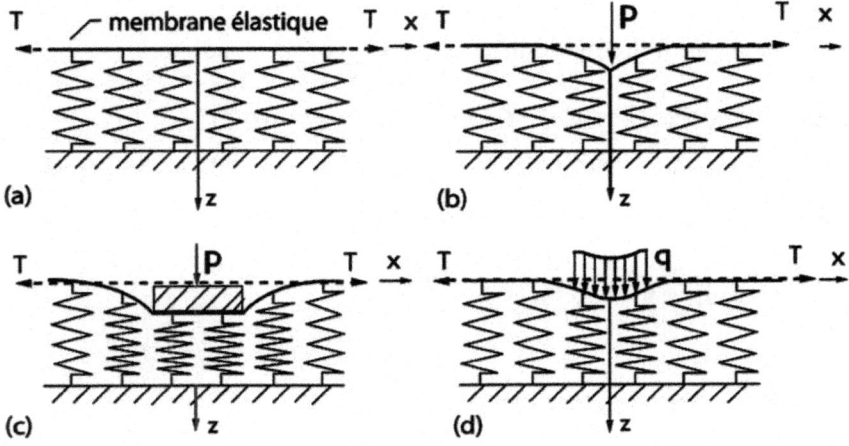

Figure 2. Model of Filonenko-Borodich [5].

with:

$$\nabla^2 = \frac{\partial^2}{\partial x^2} + \frac{\partial^2}{\partial y^2}$$

(4)

Both elastic constants of the model are the coefficient of reaction (k) and the tension (T). Figure 2shows typical examples of settlement profiles in the case of a concentrated load (Figure 2(b)), a rigid foundation (Figure 2(c)) and a flexible foundation with a uniformly distributed load (Figure 2(d)).

The study of influencing factors in this model has been the subject of several scientific publications [6].

Study of the Vertical Modulus of Subgrade Reaction (k)

Several authors have had to work on the modulus of reaction k, but all authors coming after Biot gave it a higher value than predicted by Biot [1]. In this research, the expression that was provided by Biot is used. Biot developed an empirical formula to express k [7] :

$$k = \frac{0.65 E_s}{1 - v^2} \sqrt[12]{\frac{E_s B^4}{EI}}$$

(5)

where:

E_s is the modulus of subgrade;

v_s is the Poisson's ratio of the subgrade;

B is the width of the foundation;

E_b is the Young modulus of the concrete foundation;

I is the moment of inertia of the cross section of the concrete.

Study of the Horizontal Modulus of Reaction (T)

The reaction shear modulus (T) has been proposed by Vlasov and given by the following relationship [8] :

$$T = \frac{E}{4\left(1 - v_s^2\right)\left(1 + v_s\left(1 - v_s\right)\right)} \int_0^H \Phi^2 dz$$

(6)

With:

E_s: young's modulus of the soil;

v_s: soil Poisson ratio;

H: thickness of the soil layer (depth of the rigid substrate);

$\Phi(z)$: a function which describes the variation of the displacement $w(x, y)$ along the z axis, such that:

$$\Phi(0) = 1 \text{ and } \Phi(H) = 0$$

To a relatively deep layer of soil where the normal stress may vary with depth, it is possible to use, for the function $\Phi(z)$, two types of variation (linear or nonlinear). Selvadurai [5] proposes two possible entries of $\Phi(z)$ which are given by the following equations:

$$\Phi(z) = \left(1 - \frac{z}{H}\right) \tag{7}$$

$$\Phi(Z) = \frac{\sinh\left[(H-z)\frac{\gamma}{L}\right]}{\sinh\left(\frac{\gamma H}{L}\right)} \tag{8}$$

This work shows the displacements along the two variations of $\Phi(z)$. The linear variation gives after integration, the following shear parameter:

$$T = \frac{E_s \cdot H}{12(1-v_s^2)(1+v_s(1-v_s))} \tag{9}$$

The hyperbolic variation leads to the following value of T:

$$T = \frac{E_s}{12(1-v_s^2)(1+v_s(1-v_s))} \times \frac{\left(\sin h\left(\frac{2H\gamma}{L}\right) - 2H\right)}{\sin h^2\left(\frac{H\gamma}{L}\right)} \tag{10}$$

PRESENTATION OF THE CALCULATION MODEL

To study the reinforced concrete structure, the Kirchhoff model was selected. For analysis of soil foundation, the Filonenko-Borodich model, which assimilates the soil to a spring assembly (elastic modulus k) infinitely close to each other and connected by an elastic membrane (2T voltage), was retained. The superposition of the two previous models leads to the behavior of the plate on the soil mass as shown by the Figure 3:

The theory of plate and taking into account the soil-structure interaction (biparametric model) lead to raft foundations behavioral law that may be governed by Equation (12) below:

$$D\left(\frac{\partial^4 w}{\partial x^4}+2\frac{\partial^4 w}{\partial x^2\partial y^2}+\frac{\partial^4 w}{\partial y^4}\right)-2T\left(\frac{\partial^2 w}{\partial x^2}+\frac{\partial^2 w}{\partial y^2}\right)+kw=q(x,y)$$

(12)

With the parameters D, k, T explained above.

RESOLUTION OF THE MODEL

Analytical Calculation of Displacements

In this research, it is assumed a uniform distribution of forces applied to the foundation system. Therefore q(x, y) is a constant value "Q" because it can be assumed that at any point of the foundation, there is a uniform stress distribution. For analytical resolution of the system, the double Fourier series are used. So it is assumed that q(x, y) can be written in the following form:

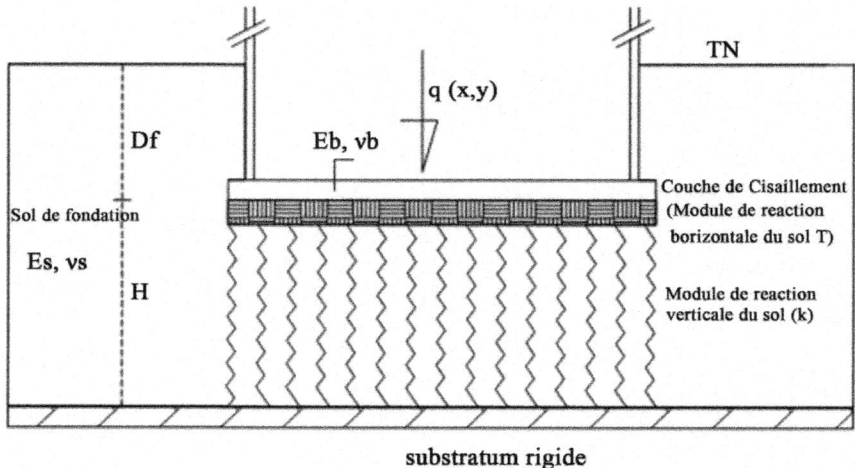

Figure 3. Mapping of the problem [6].

$$q(x,y)=Q=\sum_1^\infty\sum_1^\infty a_{mn}\sin\left(\frac{m\pi x}{L}\right)\cdot\sin\left(\frac{n\pi y}{B}\right)$$

(13)

After analytical resolution, the expression of the deflection is given as follow:

$$w(x,y) = \frac{16Q}{\pi^2} \sum_{1,3,5}^{\infty} \sum_{1,3,5}^{\infty} \frac{\sin\left(\dfrac{m\pi x}{L}\right)\cdot\sin\left(\dfrac{n\pi y}{B}\right)}{Dmn\left(\left(\dfrac{m\pi}{L}\right)^2+\left(\dfrac{n\pi}{B}\right)^2\right)^2 + 2Tmn\left(\left(\dfrac{m\pi}{L}\right)^2+\left(\dfrac{n\pi}{B}\right)^2\right)+kmn}$$

(14)

At this value of deflection, it is added the displacements of the points of the interface under the effect of the dead weight of the slab before overload. This displacement is a function of the weight of slab and the value of the elastic modulus of the subgrade and estimated at 25,000 × e/k. From the analytical solution, it is able to demonstrate the influence of the parameters of the model of behavior on the displacement of the plate. A rectangular plate of 20 m × 20 m is considered, with thickness ranging from 20 cm to 80 cm. The plate rests on a soil with elastic modulus ranging between 4 MPa and 8 MPa and Poisson›s ratio of between 0.2 and 0.45. The plate is subjected to uniform loading of 200 kN/m². The influence of all these parameters is illustrated in the following (Figures 4-16).

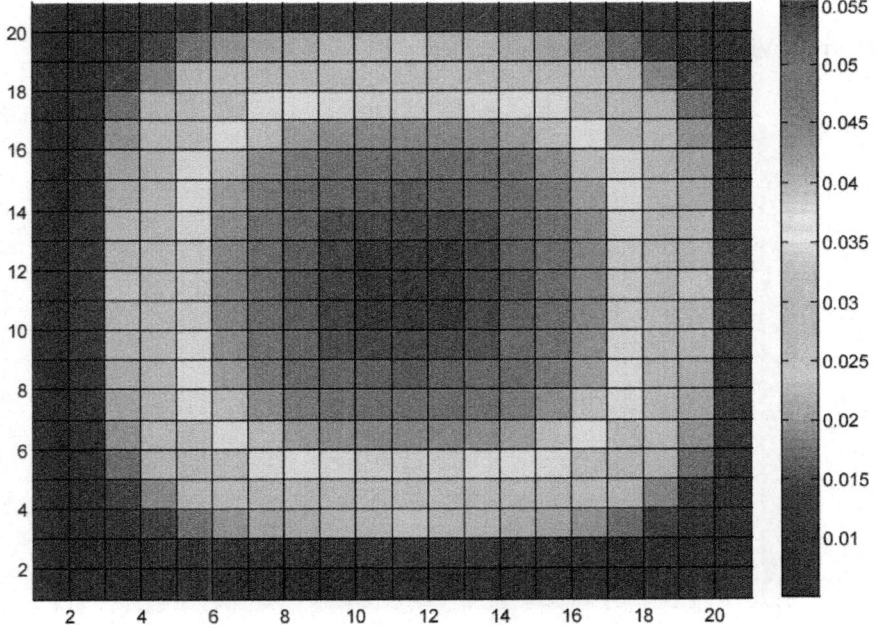

Figure 4. Visualization of 2D displacements.

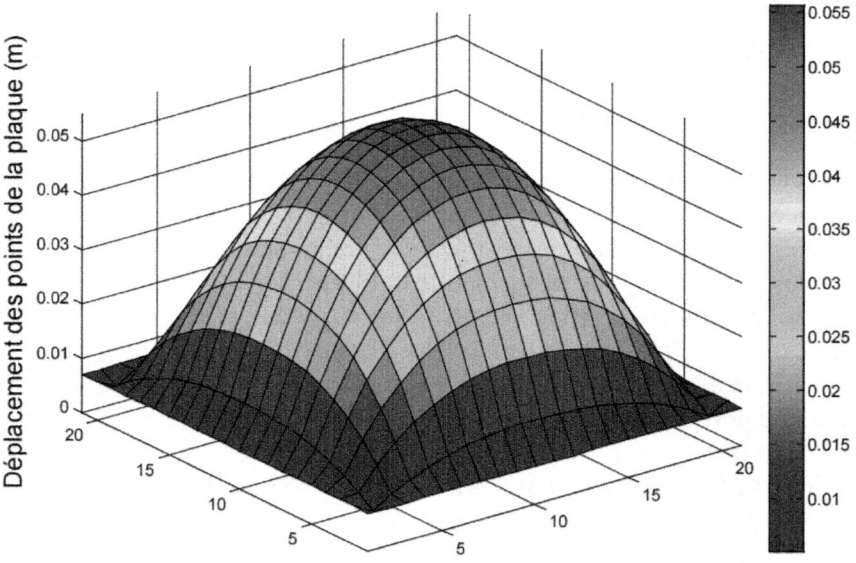

Figure 5. Visualization of 3D displacements.

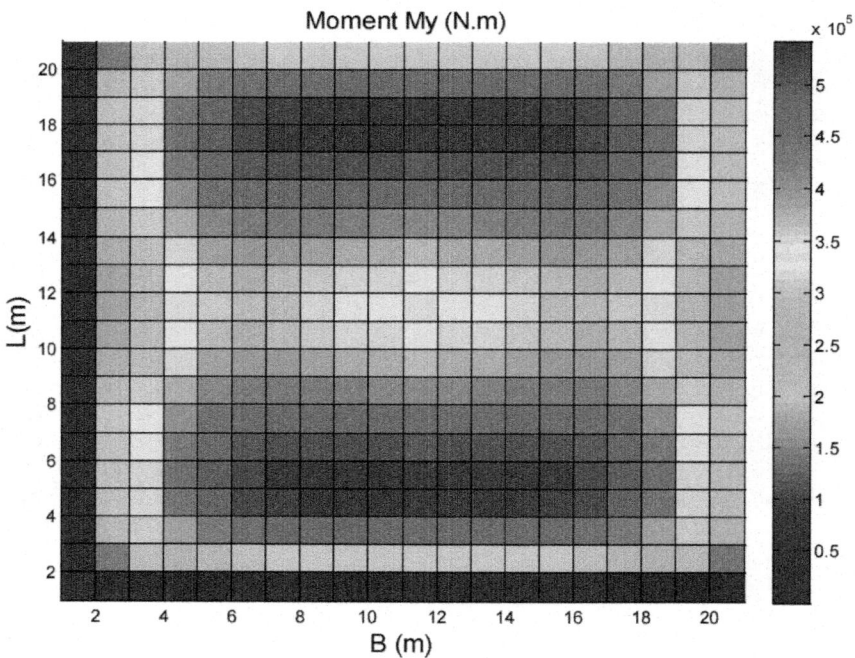

Figure 6. Bending moment M_y visualization in 2D.

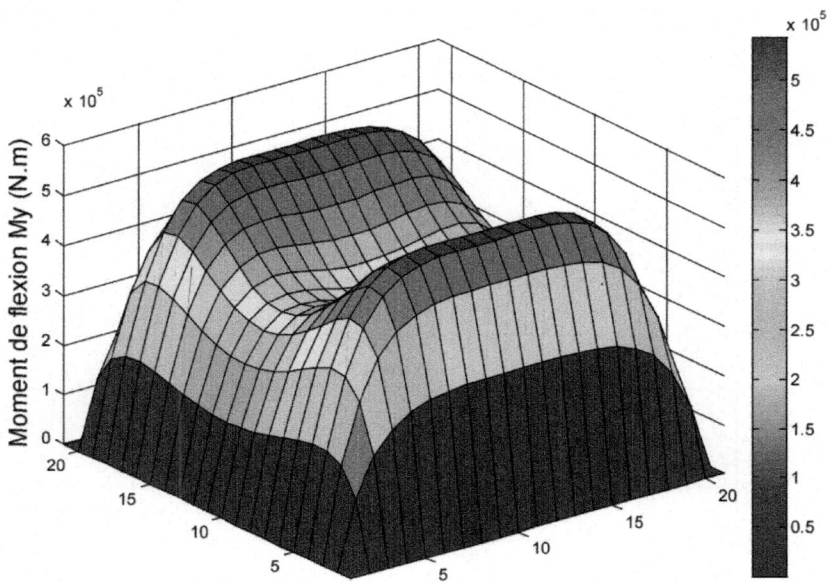

Figure 7. Bending moment M_y visualization in 3D.

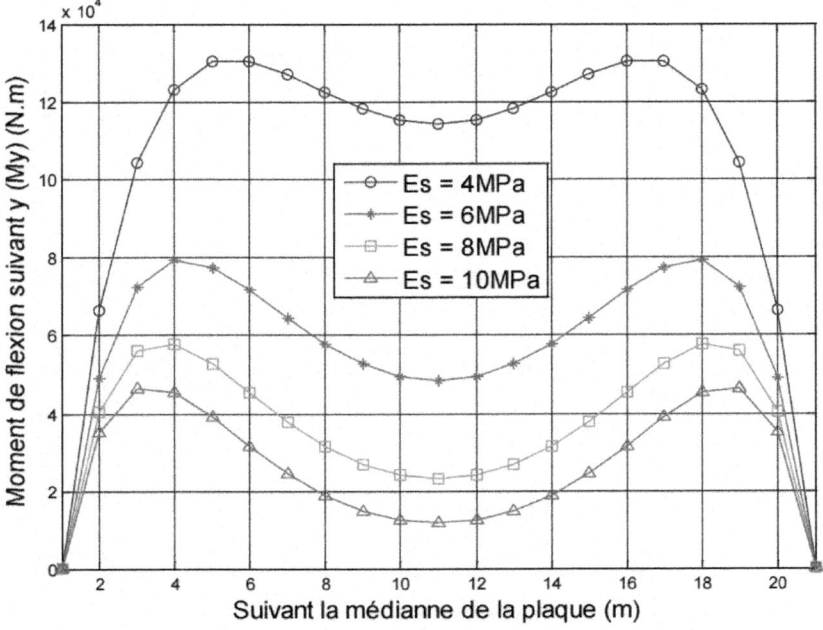

Figure 8. Bending moment M_y for various values of E_s.

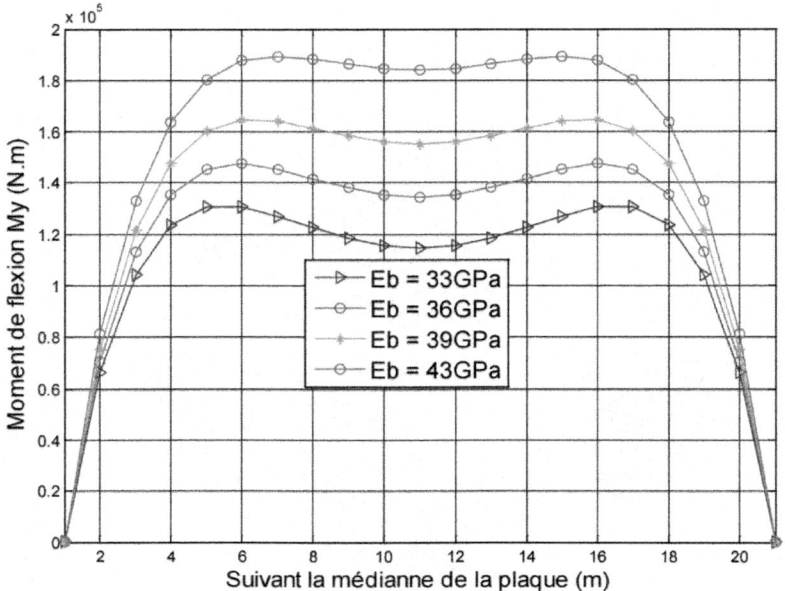

Figure 9. Bending moment M_y for various values of E_b.

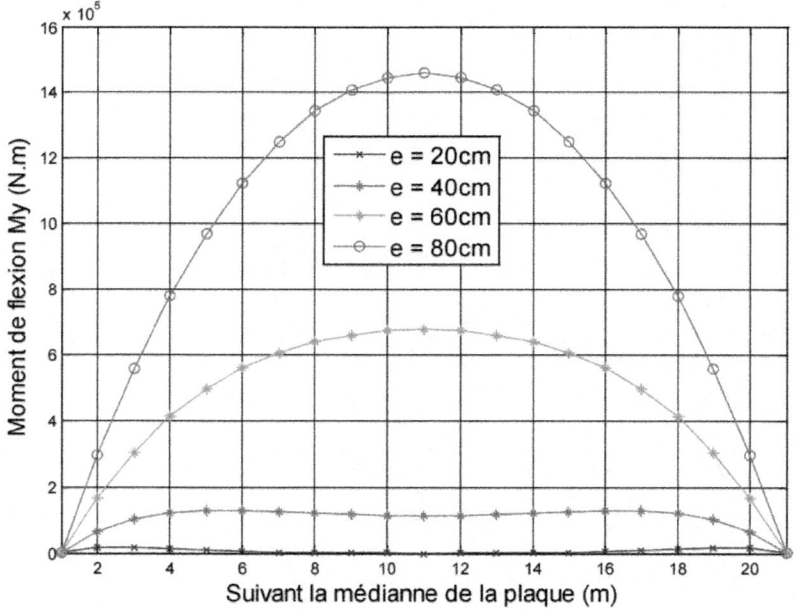

Figure 10. Bending moment M_y for various values of e.

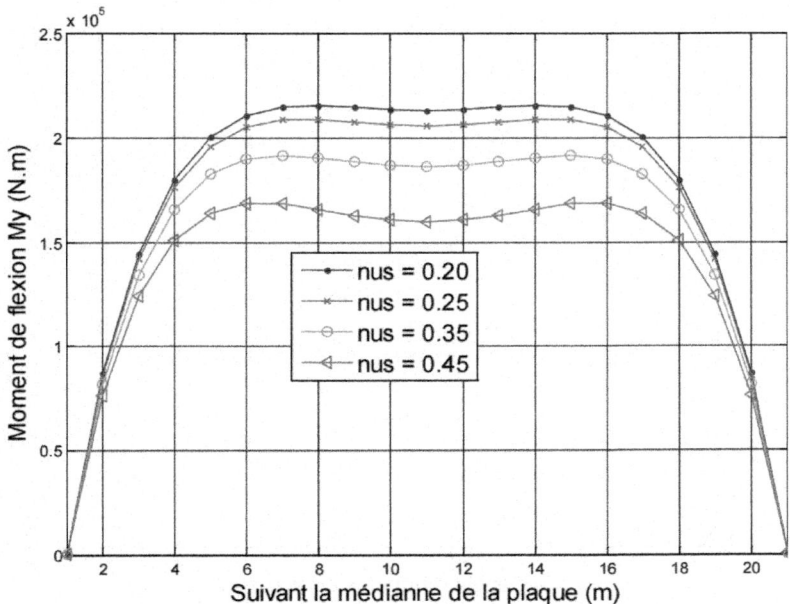

Figure 11. Bending moment M_y for various values of v_s.

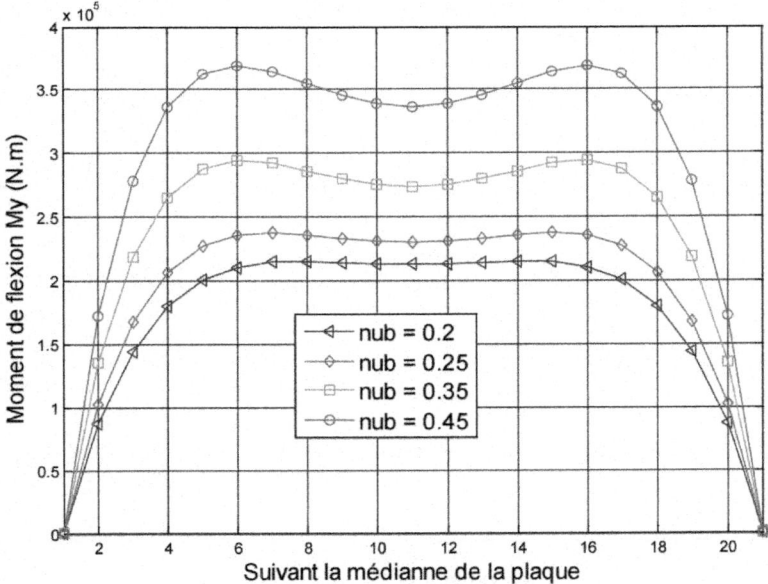

Figure 12. Bending moment M_y for various values of v_b.

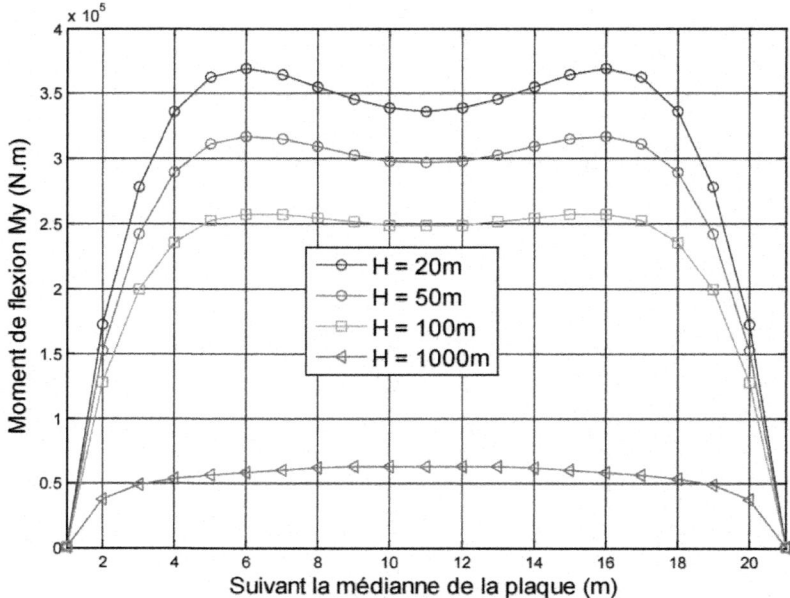

Figure 13. Bending moment M_y for various values of H.

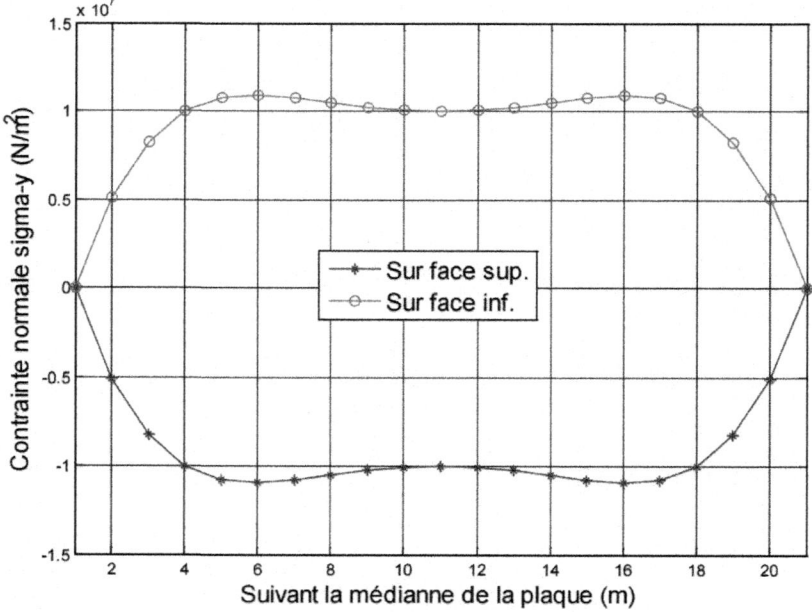

Figure 14. Normal stress σ_y on the faces.

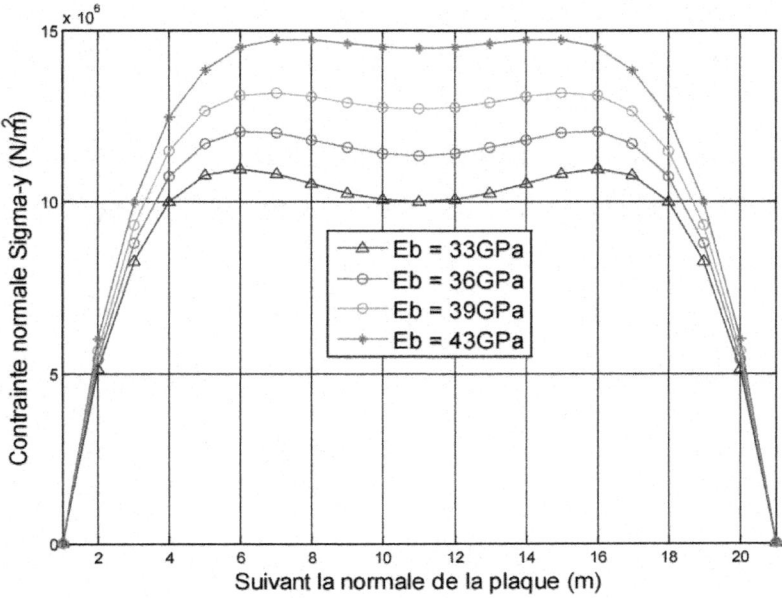

Figure 15. Normal stress σ_y for various values of E_b.

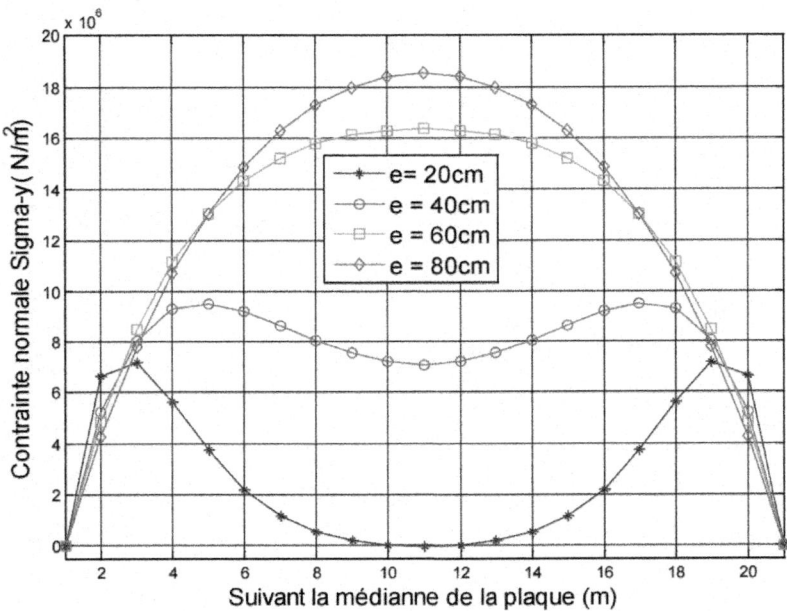

Figure 16. Normal stress σ_y for various values of e.

Calculation of Solicitations

After determining the expression of w(x, y) at any point (x, y), the solicitations are given for the following expressions.

Determination of Moments and Shear Forces

Bending moments M_x M_y and torsion moment M_{xy} are given by:

$$M_x = -D\left(\frac{\partial^2 w}{\partial x^2} + v\frac{\partial^2 w}{\partial y^2}\right)$$

(15)

$$M_y = -D\left(\frac{\partial^2 w}{\partial y^2} + v\frac{\partial^2 w}{\partial x^2}\right)$$

(16)

$$M_{xy} = -D(1-v)\frac{\partial^2 w}{\partial x \partial y}$$

(17)

Giving:

$$M_x = \frac{16QD}{\pi^2}\sum_1^\infty\sum_1^\infty \frac{\left[\left(\frac{m\pi}{L}\right)^2 + v\left(\frac{n\pi}{B}\right)^2\right]\sin\left(\frac{m\pi x}{L}\right)\cdot\sin\left(\frac{n\pi y}{B}\right)}{Dmn\left(\left(\frac{m\pi}{L}\right)^2 + \left(\frac{n\pi}{B}\right)^2\right)^2 + 2Tmn\left(\left(\frac{m\pi}{L}\right)^2 + \left(\frac{n\pi}{B}\right)^2\right) + kmn}$$

(18)

$$M_y = \frac{16QD}{\pi^2}\sum_1^\infty\sum_1^\infty \frac{\left[\left(\frac{n\pi}{B}\right)^2 + v\left(\frac{m\pi}{L}\right)^2\right]\sin\left(\frac{m\pi x}{L}\right)\cdot\sin\left(\frac{n\pi y}{B}\right)}{Dmn\left(\left(\frac{m\pi}{L}\right)^2 + \left(\frac{n\pi}{B}\right)^2\right)^2 + 2Tmn\left(\left(\frac{m\pi}{L}\right)^2 + \left(\frac{n\pi}{B}\right)^2\right) + kmn}$$

(19)

$$M_{xy} = -\frac{16QD(1-v)}{\pi^2}\sum_1^\infty\sum_1^\infty \frac{\left(\frac{m\pi}{L}\right)\left(\frac{n\pi}{B}\right)\cos\left(\frac{m\pi x}{L}\right)\cdot\cos\left(\frac{n\pi y}{B}\right)}{Dmn\left(\left(\frac{m\pi}{L}\right)^2 + \left(\frac{n\pi}{B}\right)^2\right)^2 + 2Tmn\left(\left(\frac{m\pi}{L}\right)^2 + \left(\frac{n\pi}{B}\right)^2\right) + kmn}$$

(20)

And shear forces are given by:

$$T_x = \frac{\partial M_x}{\partial x} + \frac{\partial M_{xy}}{\partial y}$$

(21)

$$T_y = \frac{\partial M_{xy}}{\partial x} + \frac{\partial M_y}{\partial y}$$

(22)

$$T_x = \frac{16QD}{\pi^2}\sum_1^\infty\sum_1^\infty \frac{\left[\left(\frac{m\pi}{L}\right)\left[\left(\frac{m\pi}{L}\right)^2 + v\left(\frac{n\pi}{B}\right)^2\right] + (1-v)\left(\frac{m\pi}{L}\right)\left(\frac{n\pi}{B}\right)^2\right]\cos\left(\frac{m\pi x}{L}\right)\cdot\sin\left(\frac{n\pi y}{B}\right)}{Dmn\left(\left(\frac{m\pi}{L}\right)^2 + \left(\frac{n\pi}{B}\right)^2\right)^2 + 2Tmn\left(\left(\frac{m\pi}{L}\right)^2 + \left(\frac{n\pi}{B}\right)^2\right) + kmn}$$

(23)

$$T_y = \frac{16QD}{\pi^2}\sum_1^\infty\sum_1^\infty \frac{\left[\left(\frac{n\pi}{B}\right)\left[\left(\frac{n\pi}{B}\right)^2 + v\left(\frac{m\pi}{L}\right)^2\right] + (1-v)\left(\frac{n\pi}{B}\right)\left(\frac{m\pi}{L}\right)^2\right]\sin\left(\frac{m\pi x}{L}\right)\cdot\cos\left(\frac{n\pi y}{B}\right)}{Dmn\left(\left(\frac{m\pi}{L}\right)^2 + \left(\frac{n\pi}{B}\right)^2\right)^2 + 2Tmn\left(\left(\frac{m\pi}{L}\right)^2 + \left(\frac{n\pi}{B}\right)^2\right) + kmn}$$

(24)

Calculation of Stresses

The stress state at the point m(x, y) is given by the following system of equations:

$$\sigma_x = -\frac{E}{1-v^2}z\left(\frac{\partial^2 w}{\partial x^2} + v\frac{\partial^2 w}{\partial y^2}\right)$$

(25)

$$\sigma_y = -\frac{E}{1-v^2}z\left(\frac{\partial^2 w}{\partial y^2} + v\frac{\partial^2 w}{\partial x^2}\right)$$

(26)

$$\tau_{xy} = -2Gz\frac{\partial^2 w}{\partial x\partial y}$$

(27)

$$\sigma_x = \frac{16Q}{\pi^2}\cdot\frac{E}{1-v^2}z\sum_1^\infty\sum_1^\infty \frac{\left[\left(\frac{m\pi}{L}\right)^2 + v\left(\frac{n\pi}{B}\right)^2\right]\sin\left(\frac{m\pi x}{L}\right)\cdot\sin\left(\frac{n\pi y}{B}\right)}{Dmn\left(\left(\frac{m\pi}{L}\right)^2 + \left(\frac{n\pi}{B}\right)^2\right)^2 + 2Tmn\left(\left(\frac{m\pi}{L}\right)^2 + \left(\frac{n\pi}{B}\right)^2\right) + kmn}$$

(28)

$$\sigma_y = \frac{16Q}{\pi^2}\cdot\frac{E}{1-v^2}z\sum_1^\infty\sum_1^\infty \frac{\left[\left(\frac{n\pi}{B}\right)^2 + v\left(\frac{m\pi}{L}\right)^2\right]\sin\left(\frac{m\pi x}{L}\right)\cdot\sin\left(\frac{n\pi y}{B}\right)}{Dmn\left(\left(\frac{m\pi}{L}\right)^2 + \left(\frac{n\pi}{B}\right)^2\right)^2 + 2Tmn\left(\left(\frac{m\pi}{L}\right)^2 + \left(\frac{n\pi}{B}\right)^2\right) + kmn}$$

(29)

$$\tau_{xy} = -\frac{16Q}{\pi^2}\cdot\frac{E}{1-v^2}z\sum_1^\infty\sum_1^\infty \frac{\left(\frac{m\pi}{L}\right)\left(\frac{n\pi}{B}\right)\cos\left(\frac{m\pi x}{L}\right)\cdot\cos\left(\frac{n\pi y}{B}\right)}{Dmn\left(\left(\frac{m\pi}{L}\right)^2 + \left(\frac{n\pi}{B}\right)^2\right)^2 + 2Tmn\left(\left(\frac{m\pi}{L}\right)^2 + \left(\frac{n\pi}{B}\right)^2\right) + kmn}$$

(30)

PRESENTATION AND ANALYSIS OF THE RESULTS

The study shows that even if the displacement is maximal at the center of the plate (Figure 4), the maximum values of the solicitations are not observed at the same point (center of the plate) as shown in Figure 6 and Figure 7. The maximum stress is located between the central third and supports of the plate. The study showed a sensitivity of stress compared to the variability of the model parameters (especially E_s as shown in Figure 8). However, it is noted a strong sensitivity of the stresses to the mechanical and geometric variability of the characteristics of the concrete (Figure 9 and Figure 11), although the movements of the model are not too sensitive to mechanical parameters of concrete (E_b, vb) as shown by [1] [8].

Figure 10 shows that the stresses are almost constants for low values of the thickness of the plate (for example e = 20 cm and e = 40 cm). The solicitations in the plate are very sensitive to the variability of the thickness. For a same load with fixed mechanical characteristics, by increasing the plate thickness, the stresses grow until the maximum value of the displacement is observed at the center of the plate (Figure 10 and Figure 16). Figure 13 shows the influence of the depth of the substratum on the internal forces and also shows that at high values of H, the solicitations are almost constant over the whole extent of the plate.

The stress distribution on the upper and lower faces of the plate (Figure 14). From this figure, it is noted a symmetrical distribution of these stresses compared with the average slip of the plate. The sensitivity of stress for various value of E_b is most felt in the center of the foundation (Figure 9 andFigure 15). The influence of soil Young's modulus (E_s) is almost felt in the same way over the entire plate (Figure 8). Figure 12 shows the variations of bending moment M_y for various values of Poisson's ratio. Figure 16 shows a variation of the normal stresses with respect to the variability of the plate thickness. For high values of the thickness (e = 60 cm and 80 cm), the maximum value of stress is observed in the center of the foundation which is the case for low values of the thickness (e = 20 cm and 40 cm).

The study also shows that even for a distributed load, the values of torsion moment and shear stress were not significant. The results of this study combined with those of Sall [1] [8] shows that calculate a foundation with one design criterion (deformation criterion or strength criterion) only may lead to bad design. This study highlights the need to check both solicitations and displacements of the foundation (soil and plate).

CONCLUSION

Results from this research show that even if the displacement is maximal at the center plate, the maximum values of the stresses are not observed at the same point. The study shows a sensitivity of solicitations compared with a variability of geometrical and mechanical characteristics of the model. Although solicitations in the plate are sensitive to mechanical properties of concrete, they are strongly influenced by the mechanical and geometrical characteristics of the soil mass. However, it should be noted that the influence of E_b is denoted in the center of the plate whereas the influence of E_s is almost the same over the entire extent of the plate. For low values of thickness, the interior efforts are almost constant on the full extent of the foundation. For high values of the depth of the rigid substrate, solicitations are almost constant over the whole of the plate. This study also shows that for the same load cases, the values of torsion moment and shear stress are not significant, compared respectively, with those of bending moments and normal stresses. This study also shows that calculations geotechnical and structural of the foundations should not be carried out separately. The results of this research should allow a good comprehension and a good taking into account of the soil-structure interaction in the calculation of foundations. In the future, it would be interesting to study the influence of the parameters of the model with more realistic laws of behavior.

REFERENCES

1.   Sall, O.A (2015) Calcul analytique et modélisation de structure en plaque interaction sol-structure en vue du calcul des fondations superficielles en forme de radier. Thèse de Doctorat de l'Université de Thiès, Thiès, 125 p.

2.   Turhan, A. (1992) A Consistent Vlasov Model for Analysis on Plates on Elastic Foundation Using the Finite Element Method. The Graduate Faculty of Texas Tech University in Partial Fulfillment of the Requirements for the Degree of Doctor.

3.   Filonenko-Borodich, M.M. (1940) Some Approximate Theories of the Elastic Foundation. Uch. Zap. Mosk. Gos, Univ. Mekh. No. 46, 3-18.

4.   Filonenko-Borodich, M.M. (1945) A Very Simple Model of an Elastic Foundation Capable of Spreading the Load Sb Tr. Mosk. Elektro. Inst. Inzh. Trans., No. 53, Transzheldorizdat.

5.   Selvadurai, A.P.S. (1979) Elastic Analysis of Soil-Foundation Interaction. Developments in Geotechnical Engineering, 17, Elsevier Scientific Publishing Company.

6.   Sall, O.A, Fall, M., Berthaud, Y., Ba, M. and Ndiaye, M. (2013) Influence of the Elastic Modulus of the Soil and Concrete Foundation on the Displacements of a Mat Foundation. Open Journal of Civil Engineering, 3, 228-233.http://dx.doi.org/10.4236/ojce.2013.34027

7.   Biot, M.A (1937) Bending of an Infinite Beam on an Elastic Foundation. Journal of Applied Physics, 12, 155-164. http://dx.doi.org/10.1063/1.1712886

8.   Vlasov, V.Z. (1949) Structural Mechanics of Thin Walled Three Dimensional System. Stroizdat, Moscow.

Chapter 9

PHYSICAL, MINERALOGICAL, AND MICROMORPHOLOGICAL PROPERITIES OF EXPANSIVE SOIL TREATED AT DIFFERENT TEMPERATURE

Jian Li[1], Xiyong Wu[1], and Long Hou[2]

[1]Department of Geosciences and Environmental Engineering, Southwest Jiaotong University, Chengdu, Sichuan 610031, China

[2]Wanzhou District Commission of Urban-Rural Development, Wanzhou, Chongqing 404000, China

ABSTRACT

Different characterizations were carried out on unheated expansive soil and samples heated at different temperature. The samples are taken from the western outskirts of Nanning of Guangxi Province, China. In the present paper, the mineral and chemical composition and several essential physical parameters of unheated expansive soil are indicated by XRD and EDX analysis. Moreover, the structural transition and change of mechanical properties of samples heated in the range of room temperature to 140°C are proved by TG-DTA and SEM observation. The mean particle diameter, density, hydraulic behaviors, and bond strength also have been investigated. The results indicate that, along with the loss of free water, physical absorbed water, and chemically bound water, the microstructure experiences some obvious change. In addition, the particle size and density both will increase rapidly before 100°C and undertake a slow growth or decline when higher than 100°C. The hydraulic behaviors and strength performance of unheated samples and the one heated at 100°C are given out as well. All these researches play fundamental role in the pollution prevention, modification, and engineering application of expansive soil.

INTRODUCTION

Expansive soils are soils that expand when water is added and shrink when it dries out. This continuous change in soil volume can cause homes and roads built on this soil to move unevenly and crack [1–3]. The special engineering properties of this soil are determined by the mineral phase and the chemical composition [4]. Therefore, researches on them are not only necessary for exploring the engineering properties of expensive soils and discussing the expansion mechanism but also indispensable as to the improvement and reinforcement of expansive soils and the discussion of new soil research techniques and methods [5, 6]. During the application of expansive soils in the construction and employment of embankment, the peculiarity is influenced by both the nature of the denudation and deliquescence of expansive soils [7–9].

The variation of temperature has a critical influence on the engineering properties of soils [10–12]. The relevant researches were as early as the 20th century AD, which were mainly focused on the soil evaporation and the invasion of precipitation [2, 13]. On the other hand, more recent studies are primarily on how temperature affects the transformation and the further strength character of soil [14–16]. De Bruyn et al. have analyzed the results of triaxial test carried out at different temperature (50°C, 80°C, and 110°C) and different confining pressure (2.1, 3.1, and 4.1 MPa), given the conclusion that the shear strength will increase with the rise of temperature [17, 18]. However, researches on the influences of temperature on the change of physical and chemical properties including microstructure and hydraulic behaviors have been rarely reported before.

In this research, the expansive soil samples collected from the western outskirts of Hanzhong of Shanxi Province were characterized to analyze the chemical and mineral components. In addition, researches on the influence of temperature on the microstructure and physiochemical properties were also carried out. It is critically important to carry out this research project for the further pollution prevention, modification, and engineering application of expansive soil.

MATERIALS AND EXPERIMENTAL PROCEDURE

Materials

Expansive soils samples were collected 1.3~1.5 m below earth surface from the western outskirts of Hanzhong of Shanxi Province, China. Approximately 3–5 Kg of soils samples was collected from six different sites. Samples were separated into several portions. Powder batches of about 500 g were kept or dried for 6 h at 20, 40, 60, 80, 90, 100, 120, and 140°C. Then the samples were removed from the furnace and cooled to room temperature in air. A powder batch of about 500 g was treated by air drying for the purpose of comparison experiment and TG-DT analysis.

Experimental Methods

X-ray diffraction (XRD) analysis was carried out on a Rigaku (Japan) D/ MAX 2500C diffractometer using GuKα radiation, voltage 40 kV, and current 200 mA, equipped with a graphite monochromator in the diffracted beam. Crystalline phases were identified using the database of the International Center for Diffraction Data-JCPDS for inorganic substances. [JCPDS, International Centre for Diffraction Data, 1601 Park Line, Swarthmore, PA, 1987.]

Thermal analysis was performed on a Netzsch (Germany) STA 449 simultaneous analyzer. Thermogravimetric (TG) and differential thermal (DT) analysis were performed in the range of 20–140°C (stripping gas: dry N^2, helium flow = 100 mL/min, and heating rate: 5°C/min). Measurements were carried out in 0.3 cm^3 volume alumina crucibles using α-alumina as reference, analyzing ≈100 mg of dry sample.

The volume frequency of particle diameter is characterized by a Winner2008A (Chinese) laser particle size analyzer, whose measuring range is 0.01–2000 μm. The density measurements were performed with a helium pycnometer (Micromeritics, Model 1305, USA). And the strength performance of expansive soil heated at different temperature is tested on a Trautwein DigiShearTM (Chinese) multifunctional direct shear test systems with the following testing condition: the shear rate is 0.03 mm/minute and the maximum shear displacement is 6.5 mm.

SEM observation EDX analysis was performed on TESCAN VEGA II scanning electron microscope for the characterization of the micromorphology of expansive soil treated at different temperature.

COMPONENTS AND PROPERTIES IN ROOM TEMPERATURE

Chemical and Mineral Components

The mineral component of expansive soil consists of clay mineral and detrital mineral. The ingredients of the clay mineral are mainly quartz, mica, feldspar, and a few of the calcites and gypsums, which are the major part of coarse grain [7]. Generally, due to the low content in expansive soil, the coarse grain has low effect on the swell-shrink property. On the contrary, the engineering properties of expansive soil are principally determined by the clay mineral, the fine grain, and especially the mineral like smectites.

The X-ray diffraction patterns of expansive soil samples are shown in Figure 1, which reflects that the main clay minerals are illite, montmorillonite, kaolinite, quartz, potash feldspar, and plagioclase according to the JCPDS cards. Moreover, the diffraction peaks have not waved and the intensity also has not changed with the variation of temperature (20, 60, 100, and 140°C), which indicates that the main mineral component has not changed. The mineral composition and the component can be given through the quantitative calculation of the intensity of the diffraction peak and full width at half maximum, which are shown in Table 1. It indicates that expansive soils from Hanzhong are mainly composed of illite and kaolin, which separately take 39% and 31% part of the total air drying sample, while the percentages of quartz and feldspar are lower than 10%. It should be mentioned that the clay mineral component is not exactly close to the soil from other sites, which is because of the different depositional environment of mother rock and rate of decay during the soil-forming process.

Table 1: The mineral components and content of expansive soil (unit: %).

Components	Montmorillonite	Illite	Kaolin	Quartz	Feldspar
Content	13	39	31	9	8

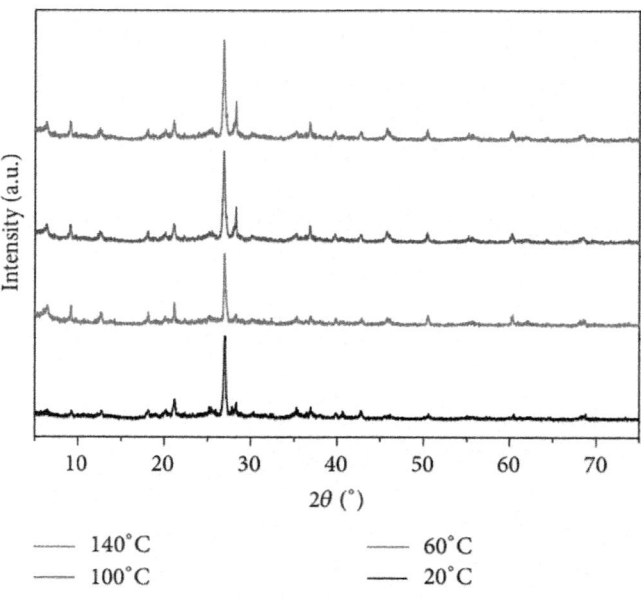

Figure 1: XRD patterns of untreated expansive soils (20°C) and samples heated at 60°C, 100°C, and 140°C.

In this project, EDX analysis was also employed to study the stability of expansive soil and the chemical composition and component which are shown in Table 2. From this table, it can be illustrated that, even though the result would vary with the EDX detection sites and the main components are SiO_2, Al_2O_3 and Fe_2O_3, which three components accounting for around 80% of expansive soil. As a consequence, the enrichment of quartz mineral in course mineral and the enrichment of aluminum silicate clay minerals in fine mineral can be concluded.

Table 2: The mineral components and content of expansive soil (unit: %).

Components	Na_2O	MgO	Al_2O_3	SiO_2	K_2O	Fe_2O_3	SiO_2/Al_2O_3
Content	3.2	6.7	19.5	47.3	8.4	12.8	2.1

Among the chemical compositions of colloidal particle of expansive soil, the molecular ratio of silicon aluminum is 3.94, which indicates that the major mineral composition is illite, corresponding to the identification result. The high components of vivacious alkali metals and alkaline-earth metals such as K, Na, Ca, and Mg demonstrate the low degree of the weathering leaching and chemical weathering and that this soil can be further weathered when the climate, aqueous medium, and oxidoreduction environment are different. Consequently, the engineering properties would be worse for the hydrophilic enhancement. The EDX pattern is shown in Figure 2 where the existence of Au element is due to the spraying for SEM observation.

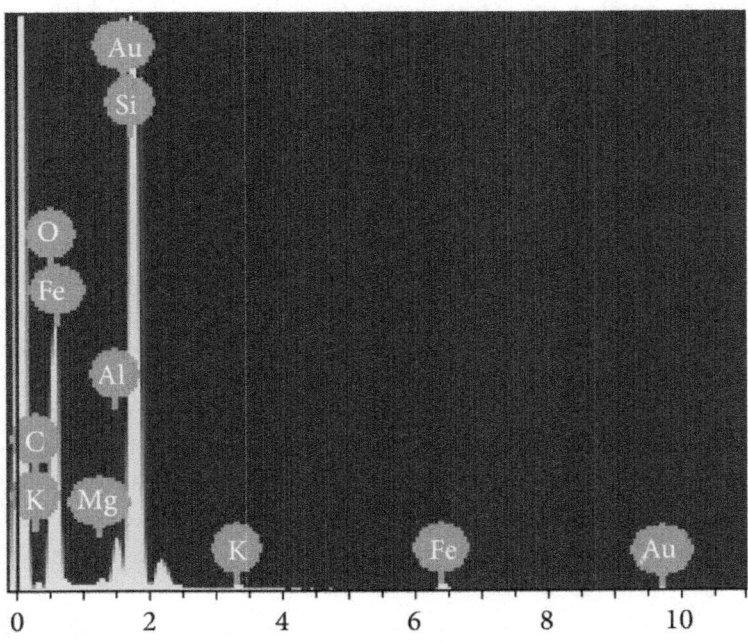

Figure 2: The result of expansive soil sample with EDX.

Physical Property

The sample soils belong to the mound hilly mudstone swell-shrinking soil area, the bed rock of which is lacustrine deposition mud and silty mud and the surface of which is soil from intense weathering and shows the structure of stratiform and color of greyish-green [19]. What is more, this soil is interbedded by silty mudstone and mudstone siltstone. That is the reason why strong expansive soils from Hanzhong are famous. The physical parameters including specific weight, liquid limit, and plastic limit of expansive soil are given in Table 3.

Table 3: The physical parameters of expansive soil.

Index	Specific weight $\gamma/kN \cdot m^{-3}$	Liquid limit $\omega_L/\%$	Plastic limit $\omega_P/\%$	Plasticity index I_P	Free swell ratio $\delta_{ef}/\%$
Soil	19.7	37.9	17.3	20.6	54

Table 3 shows that the specific weight is $19.7\gamma/kN \cdot m^{-3}$, which is similar to that of soil from the north of Hubei, China [7], while the liquid limit and plastic limit are separately 37.9% and 17.3%, in which the two can be used to identify the status of the soil and be further helpful for the application of expansive soil. In addition, the plasticity index is 20.6, which means that the expansive soil can have a high hydrophilic. Except for the parameters on the plasticity given before, another one is free swell ratio as large as 54%, which can directly reflect the high expansibility of expansive soil.

THE INFLUENCE OF TEMPERATURE

TG-DT Analysis

The TG-DTA diagram (Figure 3) shows a continuous weight loss distributed in the range of 20–140°C. The figure shows two main portions of mass loss as the rise of temperature. The first one is during the heating temperature interval of 20–100°C when the free water and some physically absorbed water are off. Before the heating temperature is up to 100°C, the sample loses 8.74% of its total weight. The proportion of physically absorbed water is small. Combined with Table 3 of physical properties, it can be known that the lost water is mainly from the free water. Therefore, the moisture content of expansive soil is around 8%, which is lower than the liquid limit and the airing treated expansive soil is in the semisolid state.

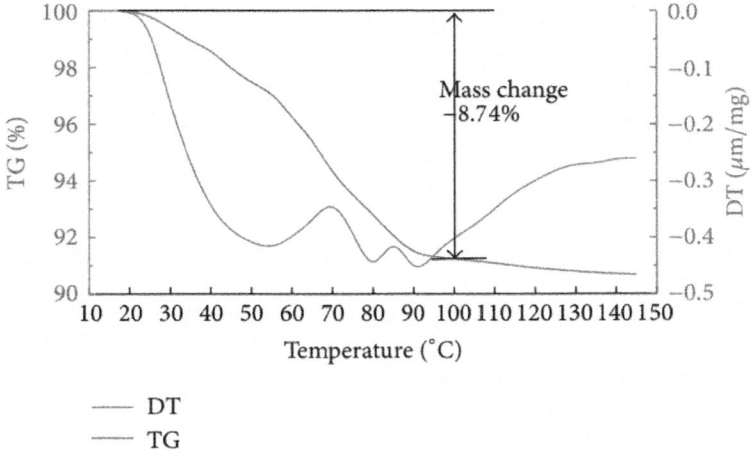

Figure 3: Thermogravimetric (TG) and differential thermal (DT) analysis diagram of heated expansive soil.

Then the mass of the sample undertakes a slight decline in the range of 100~140°C with a mass change of 1.18%. This portion of mass decrement is due to the drain of part of physically absorbed water and chemically bound water. For the existence of $Al(OH)_3$ and carbonate terrane in the clay mineral of expansive soil, the lost chemically bound water is mainly from the release of H_2O and CO_2 separately from the decomposition of $Al(OH)_3$ and carbonate terrane. The comparison of the different parts of loss illustrates that the loss of expansive soil during the heat treatment is occurring mainly before 100°C as a consequence of the elimination of free water and physical absorbed water.

Particle Size and Density

The physical properties such as particle size, density, and strength change with the increase of the heated temperature of expansive soil. These properties of expansive soil can significantly influence the occurring possibility of landslide and the settlement of the foundation [20]. The particles size distribution of the unheated expansive soil (20°C) and expansive soil heated at 100°C is shown in Figure 4(a). It can be seen that the unheated expansive soil particles are mostly in the range of 0.096–13.5 μm with a mean value of 4.67 μm. Compared with the unheated expansive soil, the heat treated one has a relatively large particle diameter. The particle diameter of expansive soil heated at 100°C is between 0.052 μm and 11.7 μm with an average value of 8.81 μm. It is believed that the rise of particle size is due to the dehydration consolidation during the evaporation process.

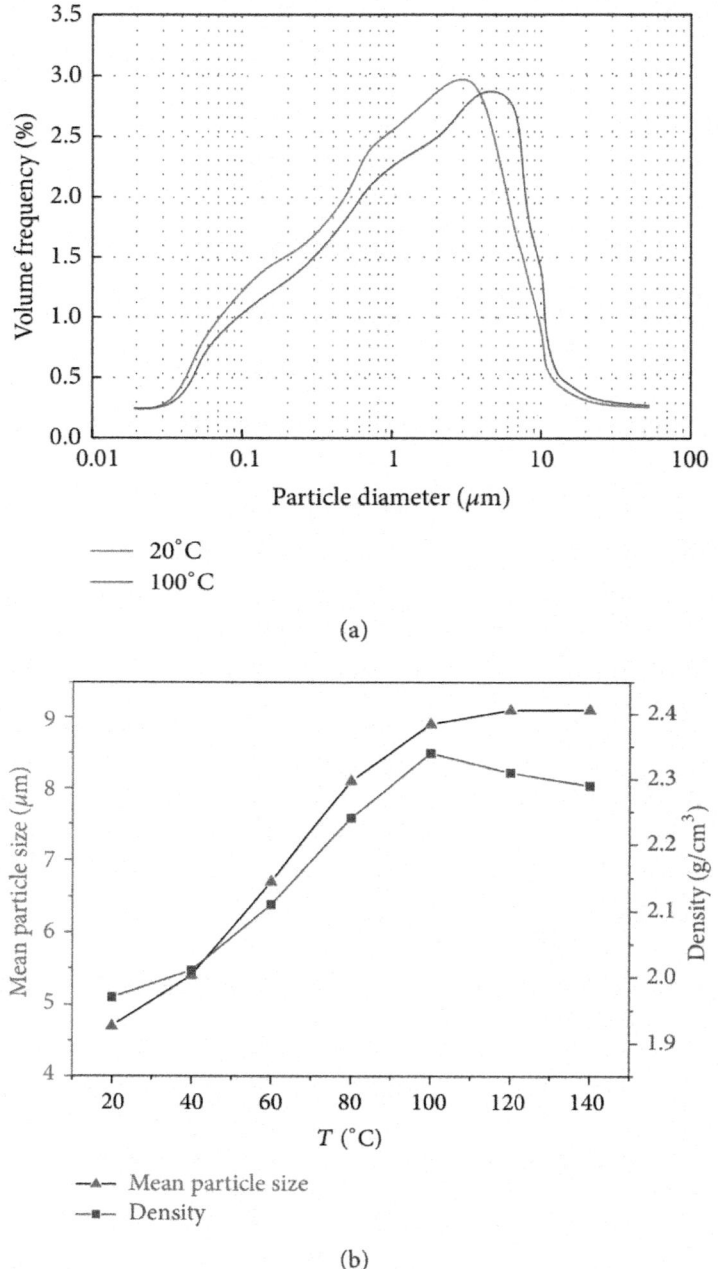

(a)

(b)

Figure 4: (a) Particle size distribution of expansive soil at room temperature (20°C) and heated at 100°C. (b) Mean particle size and density at different heated temperature.

With the change of heated temperature from 20°C to 140°C, which can indicate from Figure 4(b) that the average particle diameter of expansive soil rise from 4.67 μm at 20°C to 9.03 μm at 140°C, the temperature from 20°C to 100°C saw the rapid increase of mean particle size from 4.67 μm to 8.81 μm, followed by a slow rise to 9.03 μm at 140°C. With a similar increase tendency of mean particle size, the density of expansive soil also sharply grows from 1.94 g/cm^3 to 2.28 g/cm^3 As the temperature is increasing from 20°C to 100°C, the density of expansive soil experiences a rapid increase to the value of 2.35. And then the density declines slightly from 2.35 g/cm^3 at 100°C to 2.28 g/cm^3 at 140°C. As well as the particle size, the density also rises up because of the dehydration consolidation. As to the large growth rate before 100°C, it comes from the high water content and fast free water loss during the heat treatment. While the loss of physically absorbed water and chemically bonded water will exert a weak influence on both the density and the particle size. But the decomposition of $Al(OH)_3$ and carbonate terrane can lead to the creak and transformation of some clay mineral particle, which will further make the density at a low value.

SEM Characterization

The microstructure and the morphology play important role in the status of expansive soil and influence the expanding and shrinking behavior of expansive soil [21]. For the purpose of further comprehending the phase change progress of expansive soil during heat treatment, untreated expansive soil and samples heated at 60°C, 100°C, and 140°C are dispersed in anhydrous alcohol and grinded by ultrasonic vibration for the same time (4 h). Then the samples were observed by scanning electron microscope to obtain the micromorphology maps of these samples. The SEM images of unheated expansive soil and heat treated one are shown in Figure 5.

From Figure 5(a) it can be known that the microscopic structure of untreated expansive soil is relatively loose, with high porosity and small particle size. On the contrary, the diagrams of expansive soil heated at a series of temperatures (Figures 5(b)–5(d)) indicate that the heat treatment can improve the value of particle diameter and make the particles easy to gather together. The increasing tendency of particle size is consistent with the values measured by laser particle size analyzer as shown in Figure 4(b).

Different microstructures result from different physical and chemical progresses. With the influence of heating at 60 and 100°C, expansive soil lose the majority of its free water and part of physically absorbed water. So Figures 5(b)–5(c) present small particles and a high porosity, corresponding to a low density as shown in Figure 4(b). When heated at 140°C, with the decomposition of phases like carbonate terrane, expansive soil has lost almost all the physically absorbed water and part of the chemically bound water. So it can have larger particle and higher porosity (also higher density) than when heated at 60°C (Figure 5(d)).

(a)

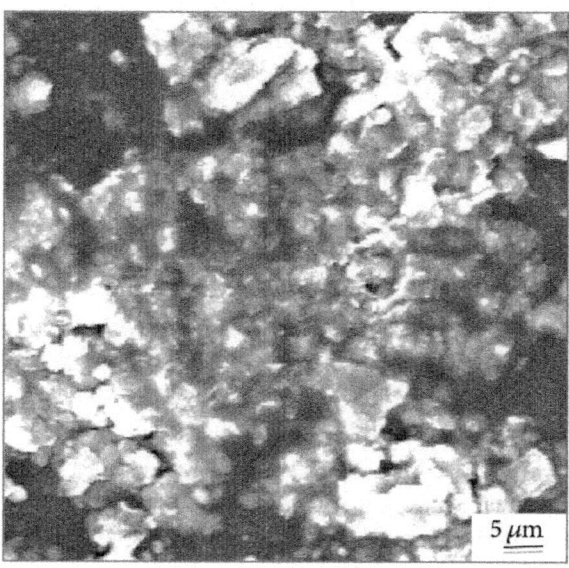

(b)

(c)

(d)

Figure 5: The SEM images of expansive soil (a) at room temperature (20°C) and heated at (b) 60°C, (c) 100°C, and (d) 140°C.

Hydraulic Behaviors and Shearing Resistance

As one type of unsaturated soil, the existence of gaseous phase, water, and soil skeleton in expansive soil is the main reason leading to the complex properties. Thus, the research on the existing form of gas phase and water phase and the migration law of gas and water under stress is indispensable to know its physical properties [20,22, 23]. And the hydraulic behaviors and strength performance play fundamental role in the pollution prevention, modification, and engineering application of expansive soil. Therefore, these relevant researches have been carried out and the test results are shown in Figure 6.

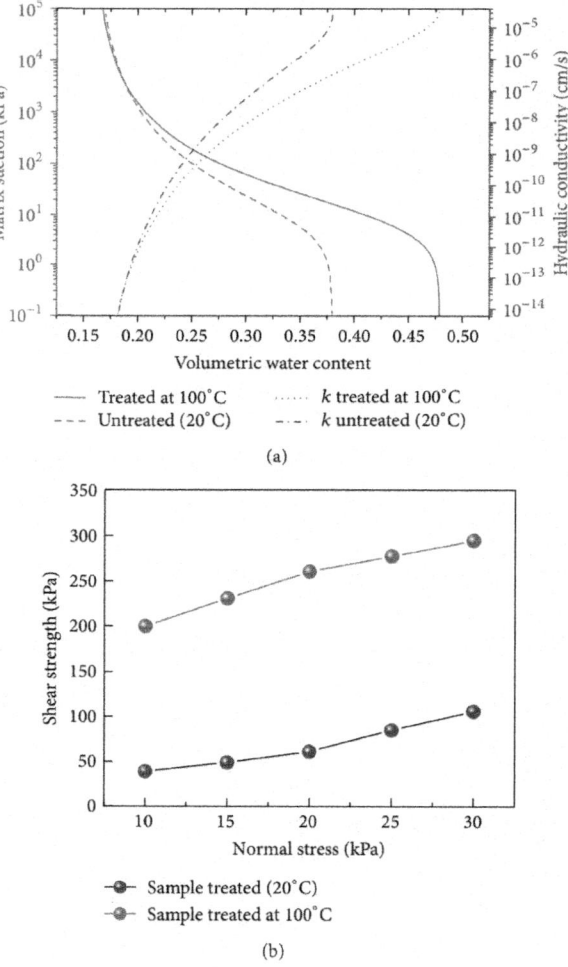

Figure 6: (a) The hydraulic characteristics of unheated (20°C) and heated (100°C) expansive soil. (b) Bond strength as the functions of heated temperature.

Both unheated expansive soil (20°C) and samples heated at 100°C for 6 h belong to alkaline engineering materials. The hydraulic behaviors of expansive soil would critically influence the spread of harmful substances in underground water and further lead to serious pollution of the surrounding soil, air, and groundwater. Therefore, a deep recognition of the hydrodynamic characteristics of expansive soil is necessary for the pollution prevention of underground water. The measured hydraulic characteristics can be illustrated by soil-water characteristic curve (SWCC) and hydraulic conductivity characteristics curve (HCF), which are shown in Figure 6(a).

The SWCC curves indicate that, with the same change of water content, the change of matrix suction of unheated expansive soil (matrix suction value is a function of several factors like free water, electric combination water, cement force, and electrochemical power) is greater than that of expansive soil heated at 100°C. This means that the water sensitivity of unheated expansive soil is greater than heated ones. As a consequence, under the same natural conditions, the stability of expansive soil after heat treatment is lower than that of unheated expansive soil. On the other hand, with the same water content, the hydraulic conductivity of the heated sample is better. It indicates that the liquid pollutants in expansive soil under high temperature are easier to filtrate and diffuse into the underground water and the surrounding environment.

The measurement shows that the strength of heated expansive soil is far higher than that of unheated soil. Then the direct shear tests were carried out for the purpose of further understanding the strength performance of the expansive soil under different temperature conditions. The testing results are shown in Figure 6(b).

The curves in Figure 6(b) indicate that, under the same experimental conditions, expansive soil heated at 100°C for 6 h has significantly high intensity. This is the same with the conclusion obtained in the sites. The bond strength of the two samples can be calculated through extending the strength envelope curve towards the left to the longitudinal axis of the coordinate system. The bond strength of heated expansive soil, 178 kPa, is significantly greater than that of unheated samples which is 26.3 kPa. In addition, it can be known that the angles of internal friction of the two materials are almost the same from the fact that two shear strength envelope curves are roughly parallel. These phenomena illustrate that the main reason why the strength of heated expansive soil is substantially higher is that it comes from the increase of bond strength.

CONCLUSION

In this project, the mineral and chemical composition and several essential physical parameters of unheated expansive soil samples taken from the western outskirts of Hanzhong of Shanxi Province, China, are tested and given out. Moreover, along with the change of temperature from 20°C to 140°C, free water, physical absorbed water, and chemically bound water are lost in sequence. The mean particle diameter undertakes a sharp increase and the SEM shows that the microstructure becomes larger and more porous due to the decomposition of phases like carbonate terrane. The hydraulic behaviors and strength performance of unheated samples and the one heated at 100°C are also given out to demonstrate the mechanical properties during the engineering application. All these researches play fundamental role in the pollution prevention, modification, and engineering application of expansive soil.

ACKNOWLEDGMENT

The authors wish to thank Yuchao Xia for providing thoughtful and critical comments on the paper. The funding for this research is provided by the National Natural Science Foundation of China (Grant no. 41172261) which is greatly appreciated.

REFERENCES

1. J. K. Mitchell, Fundamentals of Soil Behavior, John Wiley & Sons, New York, NY, USA, 1993.

2. J. R. Philip and D. A. de Vries, "Moisture movement in porous aterials under temperature gradients,"Transactions, American Geophysical Union, vol. 38, no. 2, pp. 222–232, 1957.

3. A. E. Dif and W. F. Bluemel, "Expansive soils under cyclic drying and wetting," Geotechnical Testing Journal, vol. 14, no. 1, pp. 96–102, 1991.

4. H.-Z. Yu, X.-Q. Li, and J.-W. Yao, "Experimental study and analysis of expansive soil improved with chemical medicine," Rock and Soil Mechanics, vol. 27, no. 11, pp. 1941–1944, 2006.

5. N. R. Morgenstem and J. S. Tchalenko, "Microscopic structure in kaolin subjected to direct shear,"Geotechnique, vol. 17, no. 4, pp. 309–328, 1967.

6. Q. Yang, H. Zhang, and M. Luan, "Testing study on shear strength of unsaturated expansive soils,"Chinese Journal of Rock Mechanics and Engineering, vol. 23, no. 3, pp. 420–425, 2004.

7. S.-B. Dai, J. Huang, and L. Xia, "Analysis of mineral composition and chemical components of expansive soil in North Hubei," Rock and Soil Mechanics, vol. 26, supplement 1, pp. 296–299, 2005.

8. A. A. Al-Rawas, A. W. Hago, and H. Al-Sarmi, "Effect of lime, cement and Sarooj (artificial pozzolan) on the swelling potential of an expansive soil from Oman," Building and Environment, vol. 40, no. 5, pp. 681–687, 2005.

9. B. R. Phani Kumar and R. S. Sharma, "Effect of fly ash on engineering properties of expansive soils,"Journal of Geotechnical and Geoenvironmental Engineering, vol. 130, no. 7, pp. 764–767, 2004.

10. Y. Xie, Z.-H. Chen, and G. Li, "Research of thermal effects on shear strength and deformation characteristics of unsaturated bentonite soils," Chinese Journal of Geotechnical Engineering, vol. 27, no. 9, pp. 1082–1085, 2005.

11. A. A. Basma, A. S. Al-Homoud, A. I. Husein Malkawi, and M. A. Al-Bashabsheh, "Swelling-shrinkage behavior of natural expansive clays," Applied Clay Science, vol. 11, no. 2-4, pp. 211–227, 1996.

12. Z.-H. Chen, Y. Xie, S.-G. Sun, X.-W. Fang, and G. Li, "Temperature controlled triaxial apparatus for soils and its application," Chinese Journal of Geotechnical Engineering, vol. 27, no. 8, pp. 928–933, 2005.

13. S. M. Rao and P. Shivananda, "Role of curing temperature in progress of lime-soil reactions,"Geotechnical and Geological Engineering, vol. 23, no. 1, pp. 79–85, 2005.

14. K. Harishkumar and K. Muthukkumaran, "Study on swelling soil behaviour and its improvements,"International Journal of Earth Sciences and Engineering, vol. 4, no. 6, pp. 19–25, 2011.

15. X.-W. Li, L.-W. Kong, A.-G. Guo, and Y. Zhang, "Strength characteristics of expansive soil considering effect of hydrous state," Rock and Soil Mechanics, vol. 29, no. 12, pp. 3193–3198, 2008.

16. Y. Xie, Z.-H. Chen, and G. Li, "Thermo-nonlinear model for unsaturated expansive soils," Rock and Soil Mechanics, vol. 28, no. 9, pp. 1937–1942, 2007.

17. D. de Bruyn and J.-F. Thimus, "The influence of temperature on mechanical characteristics of boom clay: the results of an initial laboratory programme," Engineering Geology, vol. 41, no. 1–4, pp. 117–126, 1996.

18. H. Xiao, K. Teng, H. Xu, et al., "Experimental study on deformation of nanning expansive soil consolidation and creep coupling," Journal of Hunan University of Technology, vol. 23, no. 5, pp. 1–6, 2009.

19. X.-D. Ou, H. Wu, and D. Zhou, "Comparative study on thermodynamics characteristics of red clay and expansive soils in Guangxi," Rock and Soil Mechanics, vol. 26, no. 7, pp. 1068–1072, 2005.

20. C. Bao, "Behavior of unsaturated soil and stability of expansive soil slope," Chinese Journal of Geotechnical Engineering, vol. 26, no. 1, pp. 1–15, 2004.

21. T. Mao and L. Xia, "Experimental research on microstructure of expansive soil in north of Hubei province," Journal of Huazhong University of Science and Technology Urban Science, vol. 27, no. 2, pp. 48–52, 2010.

22. A. T. Corey, "Meaeurement of water and air pemeability in unsaturated soil," Proceedings of the Soil Science Society of America, no. 21, pp. 7–10, 1957.

23. L. Zhan, Field and Laboratory Study of an Unsaturated Expansive Soil Associated with Rain-Induced Slope Instability, The Hong Kong University of Science and Technology (HKUST), Hong Kong, 2003.

Chapter 10

SOIL PENETRATION BY EARTHWORMS AND PLANT ROOTS—MECHANICAL ENERGETICS OF BIOTURBATION OF COMPACTED SOILS

Siul Ruiz, Dani Or, and J. Schymanski

Department of Environmental Systems Science, ETHZ, Zurich, Switzerland

ABSTRACT

We quantify mechanical processes common to soil penetration by earthworms and growing plant roots, including the energetic requirements for soil plastic displacement. The basic mechanical model considers cavity expansion into a plastic wet soil involving wedging by root tips or earthworms via cone-like penetration followed by cavity expansion due to pressurized earthworm hydroskeleton or root radial growth. The mechanical stresses and resulting soil strains determine the mechanical energy required for bioturbation under different soil hydro-mechanical conditions for a realistic range of root/earthworm geometries. Modeling results suggest that higher soil water content and reduced clay content reduce the strain energy required for soil penetration. The critical earthworm or root pressure increases with increased diameter of root or earthworm, however, results are insensitive to the cone apex (shape of the tip). The invested mechanical energy per unit length increase with increasing earthworm and plant root diameters, whereas mechanical energy per unit of displaced soil volume decreases with larger diameters. The study provides a quantitative framework for estimating energy requirements for soil penetration work done by earthworms and plant roots, and delineates intrinsic and external mechanical limits for bioturbation processes. Estimated energy requirements for earthworm biopore networks are linked to consumption of soil organic matter and suggest that earthworm populations are likely to consume a significant fraction of ecosystem net primary production to sustain their subterranean activities.

INTRODUCTION

The ability of earthworms to move and plant roots to grow through soil greatly affects their capacity to capture resources. Increasing soil compaction gradually decreases the speed and ability of biological growth [1] and movement in soil until a critical threshold is reached in some soils and movement is ceased. Prior to reaching this critical point, both earthworms and plant roots can work to break up soil, thereby ameliorating the negative impacts of soil compaction and improving their own biological habitat [2].

Bioturbation results in a network of channels that promote water infiltration and gas exchange in soil [3], and stimulate microbiological activity and preferential root growth along existing burrows. The stability of biopores is attributed to compaction at their walls [4], the excretion of biopolymers lining the burrows, and to hydrophobicity of mucilage excreted by plant roots [5] or mucus by earthworms [6].

The contributions of earthworms and plant roots to soil structural restoration after compaction [2, 7], and details of the mechanical processes involved have rarely been quantified. Capowiez and Belzunces [8] report that earthworms construct large tunnel networks with lengths ranging from 1 to 2 m per individual earthworm [8]. These burrows involve substantial amounts of displaced soil that may exceed 100 kg m^{-2} ground area per year [7]. Comparatively, plant root growth contributes less to soil biopore construction; estimates of displaced soil mass associated with plant root growth are of the order of 1 kg m^{-2} ground area per year in temperate regions [9].

Relatively few models for the mechanism of soil penetration by plant roots and earthworms have been proposed, such as the model of Greacen and Oh [10] for root growth that balances cell wall pressure and soil pressure applied on root tissue using the cone penetration analogy. This formulation became the standard root growth model ([11], [12], [13], [14]). Greacen et al. [4] have shown radiograph images of plant roots growing in a manner that compacts soil around the root circumference rather than at the forefront of the root tip. Hettiaratchi et al. [15] observed tissue thickening around the root cap that could promote soil weakening at the cap region resulting in fractures and thus lowering mechanical impedance for root growth. Misra et al. [12] have shown evidence that plant roots exert radial pressures notably larger than axial pressures with plant root radial pressures in excess of 2 MPa capable of cracking stiff rigid chalk [12, 16]. Analogously, dry compacted soil fracturing under radial pressures was measured by McKenzie and Dexter [17] in a procedure used to determine threshold radial pressures exerted by earthworms during penetration, showing that earthworms can exert radial pressures slightly above 200 kPa.

The early studies by Dexter [18] on modeling root elongation as an analogue for earthworm penetration were only recently expanded by Dorgan et al. [19] and Murphy and Dorgan [20] that proposed models for earthworm burrowing in marine sediments. Dorgan et al. [19] used crack propagation models to describe peristalsis during burrowing by earthworms in soft marine sediments. This framework has been shown to provide certain quantitative insights, however, the reliance on fracture mechanics for locomotion in wet soils or soft marine sediments appears questionable, given evidence of slow elasto-plastic deformation during root growth and expansion rates in soft soils, or considering the role of soil rheological properties that mediate other mechanical processes [21]. The primary objective of the present study is to model gradual deformation processes linked with soil penetration by earthworms. The specific objectives are to:

1. Develop a mechanical model to quantify stresses and strains associated with the soil penetration by earthworms and their dependence on soil type, hydration status, earthworm characteristics and geometry

2. Convert mechanical stress-strains to energy equivalents for different soil conditions and earthworm geometries (assuming elasto-plastic soil)

3. Relate these mechanical energy estimates to observed earthworm activity and consumption of soil organic carbon to satisfy energy needs related to soil penetration

We first present theoretical considerations necessary to develop a physically based model for the mechanics and energetics of soil penetration by earthworms that is analogous to plant root penetration models. This is followed by derivation of a simplified analytic expression for minimum earthworm pressure and energy expenditure related to creating a cavity of a given length and diameter under prescribed soil mechanical conditions. Next, we use literature values to estimate soil mechanical conditions from soil water and clay contents and generate estimates of energetic costs and physical constraints on soil penetration in a range of soil conditions. Finally, mechanical and energetic predictions are compared with experimental and empirical evidence from the literature and ecological implications are discussed.

The modeling methodology employs a continuum mechanics approach, thus assuming that penetration occurs through a homogeneous soil medium. This assumption is most applicable in compacted soil, where mechanical constraints to soil penetration are also most severe. Under many natural conditions, soils structure is likely heterogeneous and soil mechanical parameters could be anisotropic. Under these conditions, earthworms and plant roots would likely follow paths of least resistance and greatly reduce energy expenditure for soil penetration.

MATERIALS AND METHODS

The symbols used below are listed and described in Table 1.

Table 1. Table of symbols used in this study.

Symbol	Definition	SI Unit
α	Semi-Apex Insertion Angle	rad
ϵ_r	Radial Strain	$m \cdot m^{-1}$
ϵ_θ	Hoop Strain	$m \cdot m^{-1}$
ϕ	Friction Angle	rad
G	Shear Modulus	Pa
k	Shape Parameter	-
l	Distance from Cone tip	m
l_b	Burrow Length	m
v	Poisson's Ratio	$m \cdot m^{-1}$
P	Cavity Pressure	Pa
P_L	Limit Cavity Pressure	Pa
r	Radial Depth of Observation	m
R	Elasto-plastic interface	m
r_c	Cavity Radius	m
r_{c_0}	Initial Cavity Radius	m
r_f	Final Expanded Cavity Radius	m
σ_n	Normal Stress	Pa
σ_r	Radial Stress	Pa
σ_θ	Hoop Stress	Pa
σ_z	Axial stress	Pa
s_u	Soil Strength	Pa
θ_m	Water Content	$kg \cdot kg^{-1}$
θ_{min}	Residual water Content	$kg \cdot kg^{-1}$
θ_{max}	Saturated water Content	$kg \cdot kg^{-1}$
Θ	normalized water content (actual divided by saturated)	-
u	Radial Deformation	m
U	Strain Energy	J
U_0	Strain Energy Density	$J \cdot m^{-3}$

doi:10.1371/journal.pone.0128914.t001

Modeling Penetration-Expansion in Soil

The focus of the present study is on modeling the mechanics of soil penetration and the formation of macropores primarily by earthworms, in analogy to similar mechanisms applicable to elongating plant roots. We begin by discussing soil penetration by roots and worms concurrently as the equations governing the steady state mechanics of penetration-expansion for the simplified geometries are the same for both. The mechanical model formulation was first explained by Misra et al. [12], where the authors described the mechanics of axial and radial growth stresses of plant roots. A similar formulation was applied to earthworms by Mckenzie and Dexter [17]. Dexter [18], [14] unified the description of soil penetration and proposed analogy between roots and earthworms.

Despite the equivalent formulation of their mechanical soil penetration processes, earthworms and plant roots penetrate soil quite differently. Quillin [22] suggests that earthworm locomotion involves repeated penetration-expansion cycles driven by peristalsis of pressurized colonic fluid. The radial relaxation and contraction of muscles in a wave propagating manner [19, 20] results in localized radial expansion of the earthworm hydrostatic skeleton used for expanding cavities and anchoring during axial penetration. Local radial contraction is also used to elongate earthworm hydrostatic skeleton for more efficient axial penetration of soil [22] as seen in the sequence in Fig 1 (taken in soft agar). Another exclusive feature attributed to earthworms is their ability to ingest soil, however, Quillin [22] reported that earthworm burrows are generally created by displacement of soil rather than ingestion.

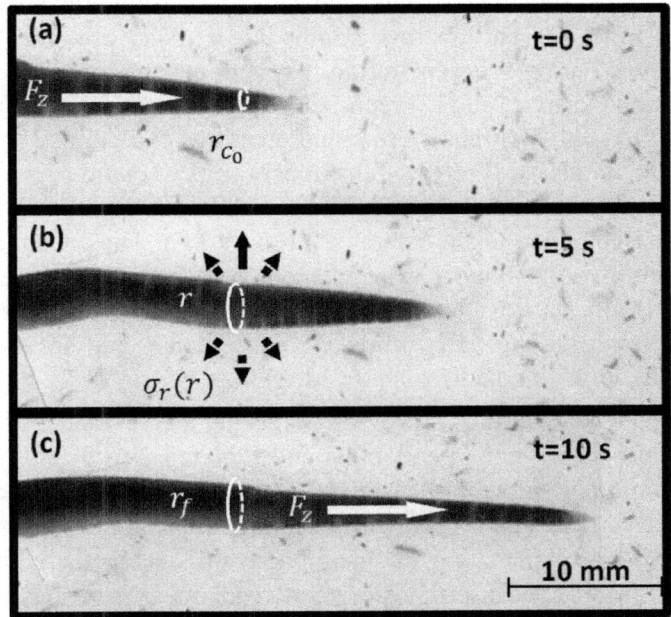

Figure 1. A sequence of images of the front segments of an earthworm moving through agar by a series of penetration-expansion steps. (a) Displays the axial penetration inducing an initial cavity. (b) Illustration of cavity expansion when collecting expanded segments. (c) Further penetration post anchoring processes.

Root penetration occurs by tissue growth rather than by peristalsis [10, 23, 24]. The resulting differences originate already at the cellular level, where earthworms utilize muscle fibers for deformation and actuation, thus enabling peristalsis. Plant cells, on the other hand, are surrounded by a stiff cell wall, having yield strengths ranging from 200 to 500 kPa [23, 24]. Root growth occurs when internal cellular pressure within the plant root exceeds the yield strength of cell wall and the soil penetration resistance. This process is facilitated by episodic reorientation of cellulose micro fibrils in cell walls during an enzymatic loosening process, and cell wall thickening during a tightening process following extension [24]. The sequence of cutting, reorientation, and rebuilding can produce pressures in excess of 1600 kPa for extended periods (up to two days) [24].

The most energetically relevant distinction between soil penetration by plant roots and earthworms relates to their penetration rates and penetration pressures. Growth-based soil penetration rates by plant roots range from 0.006 to 0.025 m day^{-1} [13, 25], which is about 20 times slower than earthworm penetration rates. Additionally, plant roots exert pressures of up to 3500 kPa [13, 15, 25–27], 100 times larger than maximal earthworm pressures [13]. Both plant roots and earthworms exhibit larger radial pressures relative to axial pressures during soil penetration, yet both extend predominantly in the axial direction. Misra et al. [12] reported axial pressures for plant roots in the range of 200 to 500 kPa relative to radial pressures exceeding 2400 kPa [13]. Keudel and Schrader [28] reported the upper limits of radial pressures exerted by endogeic earthworms in the range between 60 and 195 kPa, whereas axial pressures were limited to the range of 27 to 39 kPa. The capability of plant roots and earthworms to inhabit heavily compacted soils (Capowiez et al. [2], Dexter [18]) rely on flexible tissue, reduction of interfacial friction between their bodies and the soil (e.g., mucilage and mucus), and seeking out least mechanically impeding pathways [10, 13, 18].

Bengough and Mullins [13] reviewed mechanical resistances experienced by plant roots during cone penetration and reported 2 to 9-fold higher cone resistances compared to root resistances at similar conditions. These authors have suggested that reduced root-soil friction compared to metal-soil friction [10]could offer an explanation, but mentioned other factors including dynamic effects due to rapid cone insertion rates relative to root growth rates, and radially dominated soil deformation by roots versus axially dominated deformation by a cone penetrometer. Despite differences between penetration by plant roots and cone penetrometers, Atwell [11] stated that penetrometers provide information that correlates well with mechanical impedance experienced by plant roots. Previous models based on pressure balance of a plant root or earthworm rely heavily on cone penetration formulations. Greacen and Oh [10] utilized a formulation describing penetrometer resistance in order to compute the external pressures being applied to the plant root by the soil. More empirical methods of correlating the root growth resistance to the penetration measurements were used by Dexter [14]. Furthermore, Dexter [18] modeled soil penetration by earthworms analogously to plant roots, establishing an indirect link between earthworm penetration and cone penetration.

Motivated by Bishop et al. [29], who observed that the pressure required to produce holes in an elastic-plastic medium is proportional to the pressures required to expand a cavity to the same final diameter, we employed a common approach for these linked processes of cone penetration and cavity expansion to describe soil penetration by earthworms. The model utilizes a simple theoretical limit pressure for cavity expansion that translates to axial penetration resistance as mediated by the cone geometry. For simplicity, we consider a steady state continuum elastic-plastic mechanical model. Bishop et al. [29] found that for a sufficiently small cone angle, the resisting pressure converges towards the theoretical limiting pressure for cylindrical cavity expansion. Considering that the apex insertion angle of an earthworm and plant root is more acute than that of standardized cone penetrometers, the cylindrical cavity expansion model may be appropriate for describing the process. Bishop et al. [29] and Carter et al. [30] established the theoretical foundations for large deformation cavity expansion. Assuming a constant proportionality between the plastic and elastic fields around an expanding cavity, the authors showed that cavity pressures at large deformations would converge to a limit pressure depending on the ratio between the soil strength and shear modulus. Consequently, the stress-strain relationship obeys an associated flow law based on a Mohr-Coulomb criterion for all strains within a given radius, and act as a linear-elastic solid material outside of this boundary.

Plastic deformation is a critical process during soil penetration by earthworms in saturated and unsaturated soils. Plastic soil deformation is defined as an irreversible process resulting from cavity pressures exceeding the soil's yield threshold. Soil plasticity is crucial for the ability of earthworms to exist in soil, otherwise earthworms would constantly need to invest energy to overcome elastic rebound and would not be able to create lasting channels. Earthworms subjected to dry compacted conditions have been reported to struggle in predominantly elastic soils [17].

Cavity Expansion in an Elasto-plastic Media

At equilibrium, the stress around a cylindrical cavity decays with the radius r from the center of the cavity into the surrounding medium (soil) as:

$$\frac{\partial \sigma_r}{\partial r} + \frac{\sigma_r - \sigma_\theta}{r} = 0 \tag{1}$$

where σ_r is the radial stress, and σ_θ is the hoop (circumferential) stress (Fig 2). The mechanical transition from elastic deformation to plastic deformation is expressed by the Mohr-Coulomb criterion, relating the difference between the radial and the hoop stresses to the summation of the stresses multiplied by a component of the internal friction and the undrained soil shear strength:

$$\sigma_r - \sigma_\theta = (\sigma_r + \sigma_\theta)sin(\phi) + (2s_u)cos(\phi) \tag{2}$$

where ϕ is the internal angle of friction, and s_u is the undrained soil shear strength. Following Yu [31] and Durban and Fleck [32], we assume frictionless soils, i.e. $\phi = 0$.

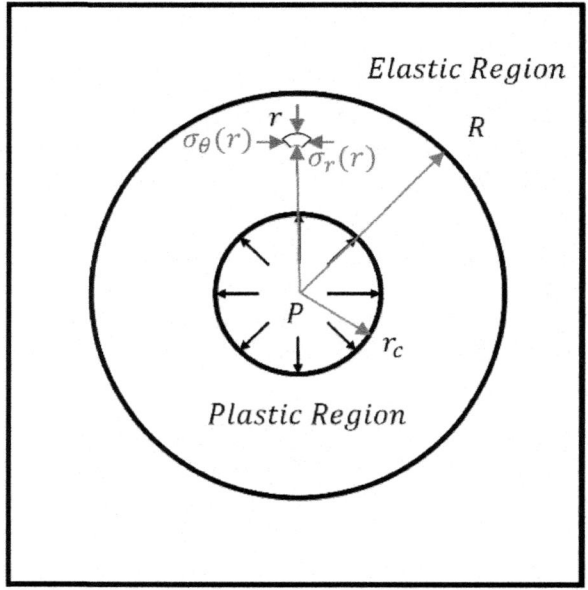

Figure 2. Concept of Elasto-Plastic cavity expansion. Cavity expansion is based on the assumption of a constant ratio between the initial cavity (r_c) and a fixed Elasto-Plastic interface (R) at a distance proportional to the internally applied cavity pressure (P). The stress field propagating into the soil, (σ_r, σ_θ) depends on the distance from the center (r).

Shames [33] provides strain-displacement relationships in axially symmetric polar coordinates. Given the magnitude of radial soil deformation u as a continuous function of the radial distance from the center of the cavity (r), the radial strain ϵ_r is defined as the deformation gradient [33 pg 536]:

$$\epsilon_r = -\frac{\partial u}{\partial r} \tag{3}$$

The hoop strain ϵ_θ is defined as the ratio between the increase in circumference $(2\pi r - 2\pi(r+u))$ over the original circumference $(2\pi r)$ [33 pg 536]:

$$\epsilon_\theta = -\frac{u}{r} \tag{4}$$

The constitutive law for a continuum material undergoing deformation resulting from an applied stress within the elastic regime is described as [33 pg 526]:

$$\epsilon_r = ((1-v)\sigma_r - v\sigma_\theta)/(2G) \tag{5a}$$

$$\epsilon_\theta = ((1-v)\sigma_\theta - v\sigma_r)/(2Gk) \tag{5b}$$

where G is the shear modulus of elasticity, v is the Poisson's ratio of the soil, and k is a shape parameter [30], distinguishing between spherical (k = 2) and cylindrical (k = 1) cavities. Elder [34] solved this problem for incompressible conditions ($v = 0.5$) considering a spherical cavity i.e. k = 2. This is equivalent to assuming that $\epsilon_r = -2\epsilon_\theta$. Adapting Elder's assumptions and applying them to a cylindrical cavity, we obtain $\epsilon_r = -\epsilon_\theta \Rightarrow \frac{\partial u}{\partial r} = -\frac{u}{r} \Rightarrow u = \frac{C_1}{r}$ (where C_1 is an integration constant). Eq 5b simplifies to

$$-2G\frac{u}{r} = -\frac{1}{2}(\sigma_r - \sigma_\theta) \tag{6}$$

Substituting $u = \frac{C_1}{r}$ in Eq 6 and subsequently Eq 6 into Eq 1 with the boundary condition that $\lim_{r \to \infty} \sigma_r = 0$ yields:

$$\sigma_r = \frac{2G}{r^2}C_1 \tag{7}$$

Substituting $C_1 = ur$ from above and solving for u gives:

$$u = \frac{\sigma_r}{2G}r \tag{8}$$

In the plastic regime, the radial stress can be estimated by substituting Eq 2 into Eq 1. For a boundary condition of $\sigma_r(r_c) = P$, where r_c represents the cavity radius, the radial stress σ_r is defined as:

$$\sigma_r = P - 2s_u ln\left(\frac{r}{r_c}\right)$$

(9)

At the elasto-plastic interface r = R, the radial stress yields the value of the undrained soil strength, $\sigma_r = s_u$, hence the deformation at the elasto-plastic interface is as follows:

$$u(r = R) = \frac{R}{2G}\left(P - 2s_u ln\left(\frac{R}{r_c}\right)\right) = R\left(\frac{s_u}{2G}\right)$$

(10)

Assuming that the change in the cavity zone $(\pi(r_c^2 - r_{c0}^2))$ equates to the change in the plastic region$(\pi(R^2-(R-u)^2))$, we substitute in for Eq 8 ($u(R) = \frac{s_u}{2G}R$) and solve for $\frac{R}{r_c}$. The relationship explicitly links cavity radius with the plastic radial domain $(r_c \le r < R)$ (Fig 2).

$$\left(\frac{R}{r_c}\right)^2 = \frac{G}{s_u}\left(\frac{1 - \left(\frac{r_{c0}}{r_c}\right)^2}{1 - \frac{s_u}{4G}}\right)$$

(11)

Inserting the relationship of Eq 11 back into Eq 10 ($\frac{R}{r_c}$), the pressure required to expand a cavity in an elasto-plastic material is expressed as:

$$P = s_u\left(1 + ln\left(\frac{G}{s_u}\left(\frac{1 - \left(\frac{r_{c0}}{r_c}\right)^2}{1 - \frac{s_u}{4G}}\right)\right)\right)$$

(12)

and under the assumption that the system is incompressible, and $\frac{G}{s_u} >> 1$, the result yields the limit cavity pressure [30] as $r_{c0} \to 0$.

$$P_L = s_u\left(1 + ln\left(\frac{G}{s_u}\right)\right)$$

(13)

Since plant root and earthworm radii are significantly smaller than standardized cone penetrometer radius, we use data from Bishop et al. [29] in conjunction with the limit pressure as a boundary condition. Considering that the deformation is predominantly plastic, we again combine Eq 1 and Eq 2 and apply the boundary condition $\sigma_r(r_b) = P_L$ to solve for the radial stress as a function of cavity size:

$$\sigma_r(r) = P_L - 2s_u ln(r/r_b)$$

(14)

where r_b (4.45 mm) is the radius used by Bishop et al. [29] to derive the expression for the cavity limit pressure P_L as described by Carter et al. [30]. This formulation facilitates calculation of the required change in cavity pressure for different radii.

Modeling Penetration Resistance

Dexter [14, 18], Bengough and Mullins [13] report that plant roots and earthworms penetrate soil in a similar manner as that of sharp penetrometers, deforming the soil cylindrically (see Fig 3). To model soil penetration by earthworms, a cavity expansion based cone penetration model is employed. Yu [31] and Durban and Fleck [32] developed a semi-analytic expression based on cavity expansion for rough and smooth penetration at different angles. This formulation considers the angular effects when neglecting friction. The penetration resistance stress can be expressed as [31, 32]:

$$\sigma_z = s_u \left(\pi + 2\alpha + \sin^{-1}(m) + \frac{D}{2} + m \cot \alpha - \sqrt{1 - m^2} - 1 \right) + \sigma_r \tag{15}$$

where α is the semi-apex cone insertion angle, $m \in [0, 1]$ is the gauge of roughness where $m = 0$ is lubricated, and $m = 1$ is rough, and

$$D = \frac{\sin\left(\frac{\pi - \alpha}{2}\right) + m \sin(\pi - \alpha)}{\cos\left(\frac{\pi - \alpha}{2}\right) - \cos(\pi - \alpha)}$$

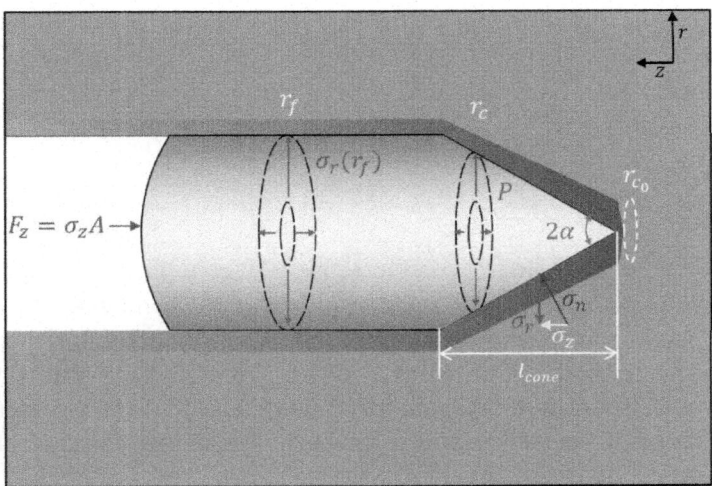

Figure 3. Cylindrical cavity expansion sequentially determines steady state penetration of acute cones. The conical cross section applies a boundary pressure that opens a cavity to some final steady state cylindrical burrow.

The penetration resistance force can be determined by integrating the total penetration resistance stress σ_z along the length of the cone for the changing cross sectional area:

$$F_z = \begin{cases} 2\pi \int_0^{r_f} r\sigma_z dr, & 0 \leq l < l_{cone} \\ \pi r_f^2 \sigma_z, & l > l_{cone} \end{cases} \tag{16}$$

where r_f is the cone base radius, l_{cone} is the cone length, and l is the penetration depth. For $l \leq l_{cone}$, $l = r\cot(\alpha)$ (see Fig 3). For sufficiently long penetration depth where $l >> l_{cone}$, the second expression suffices to account for the effective penetration force.

Strain Energy for Cavity Expansion in Plastic Soil—Ecological Considerations

All energetic costs calculated throughout the text pertain to the mechanical energy required for soil penetration by earthworms (involving plastic deformation and displacement of the soil). The mechanical strain energy is defined as the amount of energy invested to induce deformation based on the following:

$$U = \int_{z_0}^{z_f} F_z dz \tag{17}$$

where U is the strain energy of the system, F_z is the penetration resistance force, and dz is the change in the penetration depth. By substituting Eq 16 into Eq 17, we obtain:

$$U = \begin{cases} 2\pi\cot(\alpha) \int_0^{r_f} \int_0^r \xi\sigma_z d\xi dr, & 0 \leq l < l_{cone} \\ \pi r_f^2 \sigma_z l, & l > l_{cone} \end{cases} \tag{18}$$

where σ_z is the axial stress, α is the semi-apex cone insertion angle, l is the axial depth of penetration, l_{cone} is the axial length of the cone tip, r is the cavity radius, r_f is the earthworm base radius, and ξ is a dummy variable used for integration, representative of radius. For $l >> l_{cone}$, the first term (during insertion) only accounts for a marginal amount of the total strain energy, therefore the second term can be used to approximate the total strain energy related to a penetration depth l.

The maximum strain energy density is estimated by dividing the strain energy required to construct small earthworm burrows by their burrow volume, resulting in the following expression:

$$U_0 = \frac{U(r_{min})}{\pi r_{min}^2 l}$$

(19)

where r_{min} is taken as the smallest earthworm radius based on ecological parameters, $U(r_{min})$ is the strain energy required for a small cavity, $pr^2 min^l$ is the resulting cavity volume. This expression for the strain energy density is used to determine the minimum amount of soil organic carbon content that would meet the mechanic energy demand of earthworm motility.

Soil Mechanical and Biophysical Properties

The mechanical properties of soils, and mechanical constraints on bioturbation are functions of soil water content, porosity, clay minerals, and other parameters. We explored a range of these soil physical parameters to delineate mechanical stresses and energy costs related to soil penetration by earthworms. A range of data for soil shear strength and elasticity as functions of clay and water contents were obtained from the literature (Table 2). The soil modulus of elasticity was estimated from the clay mechanical properties [38]. To capture the dependence of shear modulus of clay on its water content (Fig 4), these results were expressed in terms of water content and cohesion [39], suggesting that the inferred shear moduli were derived from samples with relatively low water contents. Studies conducted on saturated clays show greatly reduced shear modulus and soil shear strength [21].

Table 2. Soil mechanical parameters used in cavity expansion simulations, and their literature sources. θ_m: water content, G: shear modulus, s_u: soil strength.

Clay content %	$\theta_m[kg\ kg^{-1}]$	G[kPa]	$s_u[kPa]$
15–25	0.20	200 [21]	2–20 [21, 35, 36]
15–25	0.25	70 [21]	0.6 [21]
15–25	0.35	-	5 [35]
15–25	0.40	15 [21]	0.20–20 [21, 35, 36]
40–50	0.20	500 [21]	-
40–50	0.45	150 [21]	4 [21]
40–50	0.70	50 [21]	2 [21]
100	-	5000–17,000 [37]	50–100 [37]
100	0.20	4000 [38]	-
100	0.25	3000 [38]	-
100	0.30	2000 [38]	6 [39]
100	0.40	500 [21]	4 [21]
100	0.50	350 [21]	1.5 [21]

doi:10.1371/journal.pone.0128914.t002

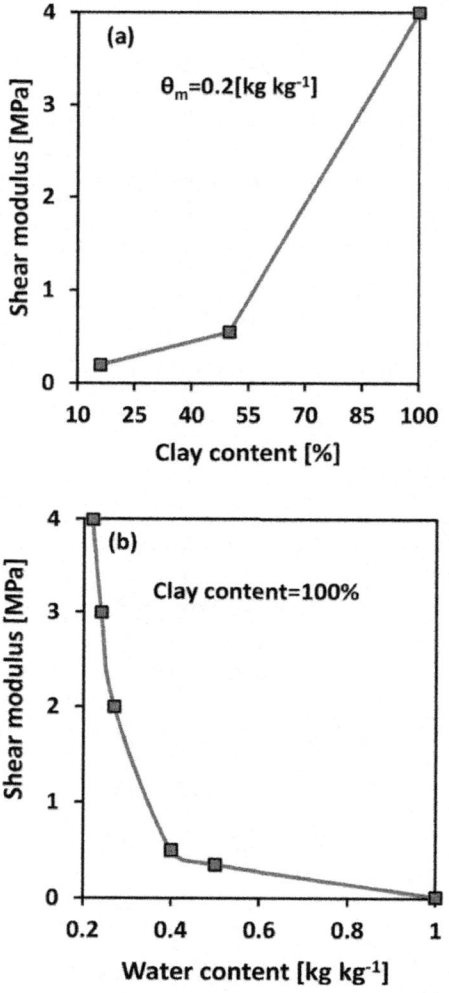

Figure 4. Measured shear modulus values. (a) Different clay contents for fixed water content of 0.2 kg kg^{-1} [21]; (b) different water contents for 100% clay content [21, 38, 39].

Earthworm physical and ecological parameters were used to determine the mechanical limitations and to estimate the strain energy requirements for soil penetration by earthworms. Observed values of earthworm pressure thresholds were obtained from Newell [40], Keudel and Schrader [28], and McKenzie and Dexter [17], providing insights into the range of soil hydrological conditions that permit earthworms to penetrate soil. The internal earthworm pressure values rarely exceed 230 kPa, with mean values in the range of 60 to 200 kPa [17, 28, 40]. Estimates of earthworm radii were obtained from Vandenbygaart et al. [41] and Ehlers [42] with values ranging between 1.0 and 5.5 mm [41], and an earthworm population mean radius of 2.5 mm [42]. Estimates of tunnel lengths were obtained from Capowiez and Belzunces [8]. To estimate annual mechanical energy requirements per unit soil area (or volume) for a typical earthworm community, knowledge of population density and annual penetration rates were needed. Capowiez and Belzunces [8] reported penetration rates for individual earthworms in the range of 0.1 to 0.2 m day^{-1} for measurements over 200 hours. Earthworm population density values were obtained from Daniel [43] and Chan [44] measured at 200 mm soil depths, Fragoso and Lavelle [45], measured at 100 mm soil depths [44] (data are presented in Table 3).

Table 3. Earthworm physical parameters. r_f: worm radius; l_b: tunnel length; $\frac{dl_b}{dt}$: penetration rate; n: population density; P: pressure.

Classification	r_f[mm]	l_b[mm]	$\frac{dl_b}{dt}$ [$\frac{m}{day}$]	n[ind m^{-3}]	P[kPa]
Swiss Meadows [43]	-	-	-	700 – 1550	-
Tropical Forest [45]	-	-	-	40 – 4000	-
Agricultural Fields [44]	1 – 5.5 [41, 42]	-	-	300 – 700	-
Lab	-	1000 – 1600 [8]	0.12 – 0.20 [8]	-	3 – 230 [17, 49]

doi:10.1371/journal.pone.0128914.t003

The analysis of penetration expansion was based on models that provided information for determining the amount of mechanical energy required to expand a cavity to radius r_f. The range of radii was based on the smallest and the largest earthworm radius in Table 3 (r_f = 1 to 5 mm). Standard mechanical soil properties and their connection to hydration status and clay content were derived from the relations in Ghezzehei and Or [21] (see Table 4). For better comparability between soil types, the hydration status was expressed as normalized water content, defined as

$$\Theta = \frac{\theta - \theta_{min}}{\theta_{max} - \theta_{min}} \tag{20}$$

where θ_{max} is the highest reported (saturated if available) water content, θ_{min} is the lowest reported (residual if available) water content, and θ is the actual gravimetric water content. The range of reported water content values are tabulated in Table 4. To account for lubrication by biopolymers (e.g. earthworm mucus), the interface between the soil and earthworm was considered to be a frictionless and smooth boundary. The range of apex angles representing earthworm geometries ranged from α = 1 to 45°.

Table 4. Input parameters for the Mechanical cavity expansion simulation. θ_{min}: residual water content; θ_{max}: saturated water content; Θ: normalized water content; G: Shear modulus; s_u: soil strength. (Values marked with an asterisk (*) were extrapolated based on the trend lines presented in [21, 46]).

Clay	θ_{min}	θ_{max}	Θ	G	s_u
[%]	[kg/kg]	[kg/kg]	[-]	[kPa]	[kPa]
16	0.045	0.4			
			1	15	0.17
			0.44	200	2.30
			0.1*	2386*	40*
50	0.1	0.85			
			1	10	0.6
			0.47	150	4
			0.15*	5668*	38*

doi:10.1371/journal.pone.0128914.t004

From Penetration Mechanics to Soil Displacement Energy Estimations

Using strain energy per unit length derived from mechanical considerations, estimates of energy demands for bioturbation were derived for different ecological conditions. Using the values of l_b as the length of earthworm channels from Table 3, we calculated mechanical energy costs of soil penetration (per length) and determined the amount of strain energy required for incremental soil penetration. In conjunction with data regarding earthworm penetration rates and earthworm population densities in various ecological systems (Table 3), we estimated the amount of mechanical energy transferred to the soil by a population of earthworms. For simplicity, we consolidated reported vertical distributions of earthworm radius to a single mean value of 2.5 mm, assuming that the value does not vary too much in the top half meter of soil [42]. The estimated energetic costs of soil penetration were translated to equivalent soil organic carbon consumption by earthworms, using a conversion coeefficient of 0.0484 g_{carbon} J^{-1} [7].

Model Evaluation—Comparison to Numerical and Empirical Data

Predictions by the analytical model of cavity expansion (Eq 12) were first evaluated in comparison with numerical calculations for the same conditions, and then with experimental results. A plane strain steady state finite elements model (FEM) was constructed using COMSOL [47]. The finite element model simulated deformation driven cavity expansion in an incompressible elastic-perfectly plastic medium with boundary walls infinitely far from the expanding cavity. Both the finite element model and the analytic model were compared for 9-fold expansion from an intial radius.

Additional comparisons were performed with an explicit cone penetration model developed by Walker and Yu [48]. Their model uses an adaptive finite element remeshing algorithm in order to simulate the actual motion and geometry of soil penetration directly rather then with cavity expansion. The analytic solution expressed in Eq 15 was used for a cone roughness of m = 1 and a semi apex angle of α = 30°. Both models assume s_u = 10 kPa and G = 1000 kPa (taken from Walker and Yu [48]).

The model was lastly tested against experimental penetration data of Kurup et al. [49], where two replicates of a silty clay (50% Kaolinite and 50% Edgar fine sand) were measured to have soil strengths and shear moduli of a) s_u = 65 kPa and G = 567 × s_u and b) s_u = 40 kPa and G = 150 × s_u under saturated conditions. We assumed a friction effect of m = 1, semi apex angle of 30°, and utilized the final radial stresses for cone radii of 5.64 and 6.36 mm associated with the cone designs of the miniature piezocone penetrometer and minature quasi-static cone penetrometer respectively [49].

Comparison to Crack Propagation Model

Results from the present model were also compared with data provided by Dorgan et al. [50], who assumed that earthworms penetrate sediments by crack propagation and estimated energy requirements for soil penetration using linear elastic fracture mechanics. Estimations of energy requirements for soil penetration by crack propagation were conducted using linear elastic fracture mechanics (LEFM) utilizing an energy formulation defined by Dorgan et al. [50]:

$$U = K_{Ic}^2 \frac{(1 - v)}{2G} lr_f$$

(21)

where K_{Ic} is the fracture toughness, v is the Poisson's ratio, G is the shear modulus, l is considered to be the distance over which the crack grows, and r_f is considered to be the width of the crack. For simplicity, this study assumes that the width of the crack is the same size as the radius of the earthworm penetrating the soil, and the length that the crack grows is equal to the depth of a given earthworm tunnel. For comparison with our model, we assumed a 1 m long penetration depth with a radius of 1.2 mm [50].

The fracture toughness changes as a function of soil water content. Values for fracture toughness for lower water contents were collected from Hanson et al. [51] and Wang et al. [52]. For a soil with clay content of 15–25% and saturated conditions (water content of $\theta_m = 0.44$ kg kg^{-1}), the mechanical shear modulus is equivalent to that of gelatine used in the study by Dorgan et al. [50] (G = 1.4 kPa). The fracture toughness parameters were fit to a continuous curve plotted against water content in order to estimate mechanical energy investments for a wider range of water contents (seen in Fig 5).

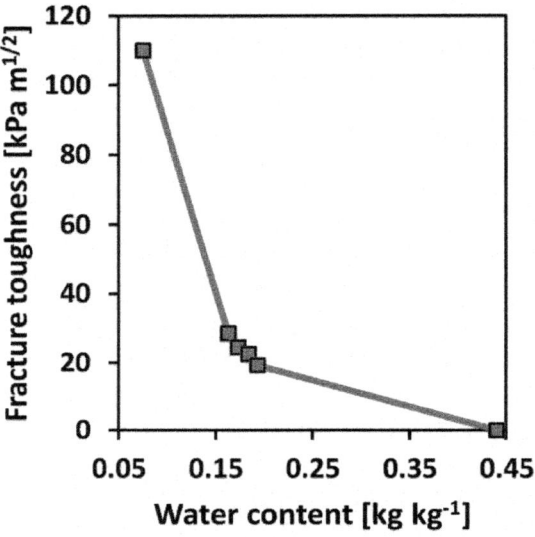

Figure 5. Fracture toughness vs water content. [50–52]. Continuous curve was plotted through the data points in order to approximate fracture toughness values at different water contents.

RESULTS

Model Evaluation—Comparison to Numerical and Empirical Data

The results depicted in Fig 6 illustrate that the limit pressures predicted by the analytic model and the finite element COMSOL model converge for large final cavity radii compared to initial radii, given same soil strength and shear modulus. Note that PP_L^{-1} is the ratio of cavity pressure to limit pressure, and the limit pressure is invariant with respect to cavity radius [30].

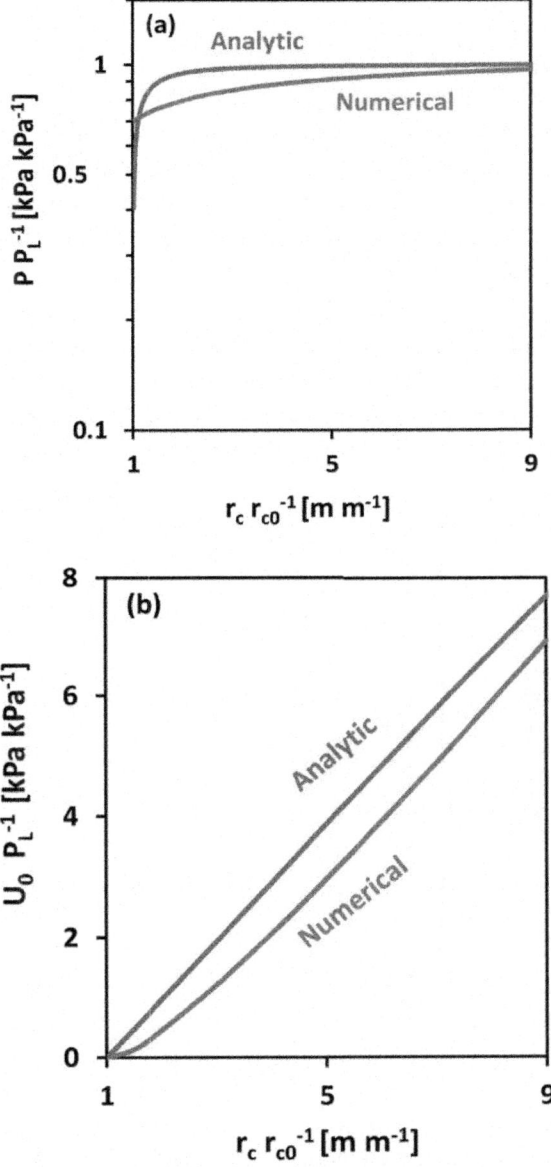

Figure 6. Analytical cavity expansion model vs. Finite Element cavity expansion model. (a) Relative cavity pressure vs. relative cavity radius; (b) strain energy density scaled by limit pressure vs. relative cavity radius scaled by initial radius. For both models, changes in soil mechanical parameters only changed the magnitude of P_L. Both analytic and numerical models showed close magnitudes of P_L. The discrepancy between the strain energy density values was measured as:

$$\frac{\left\| U_{0,NUM} - U_{0,ANA} \right\|_\infty}{\left\| U_{0,NUM} \right\|_\infty} = 0.11.$$

The results show that the convergence to the limit pressure occurs at a larger radial deformation in the numerical solution than predicted by the analytic solution. However, the difference between strain energy densities for the analytic solution and the numerical solution were less than 10% at 9-fold deformation (Fig 6 (b)).

The penetration stress versus radial strain was compared between the cavity expansion based cone penetration model and the geometrically explicit adaptive FEM cone penetration model developed by Walker and Yu [48]. The comparison suggests that the simplifying assumptions employed in the analytical model introduced a relative error of less than 20% (Fig 7).

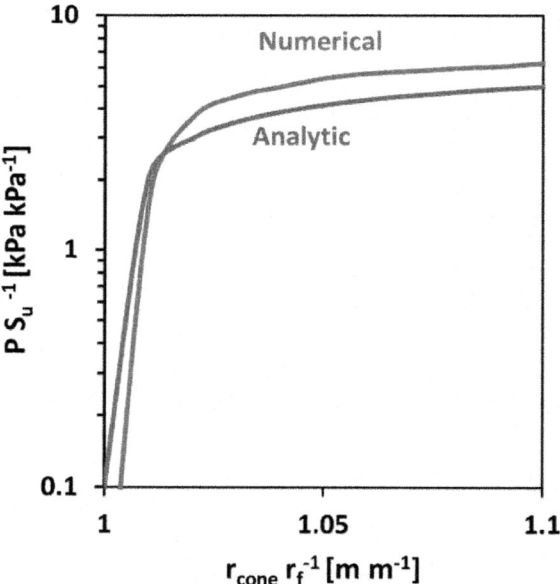

Figure 7. Comparison of analytical cavity expansion based model with an adaptive finite element explicit penetration model (Walker and Yu [48]). Comparison is drawn between the relative penetration stress vs the radial strain. Penetration stress is scaled by the shear soil strength. The discrepancy between the pressure values over two orders of magnitude never exceeds

$$\frac{\left\| P_{NUM} - P_{ANA} \right\|_\infty}{\left\| P_{NUM} \right\|_\infty} < 0.2.$$

Finally, the soil impedance predicted with the present penetration-expansion model was compared with experimental impedance results reported by Kurup et al. [49] down to a soil depth of 150 mm (seen in Fig 8). The comparison of the simulated stresses and the experimental data reveals an error between measured and predicted soil mechanical impedance of about 8–18% for the first test, and 20 to 35% for the second test. The discrepancy towards the tip could be related to dynamic effects that are not properly accounted for in the current steady state solution.

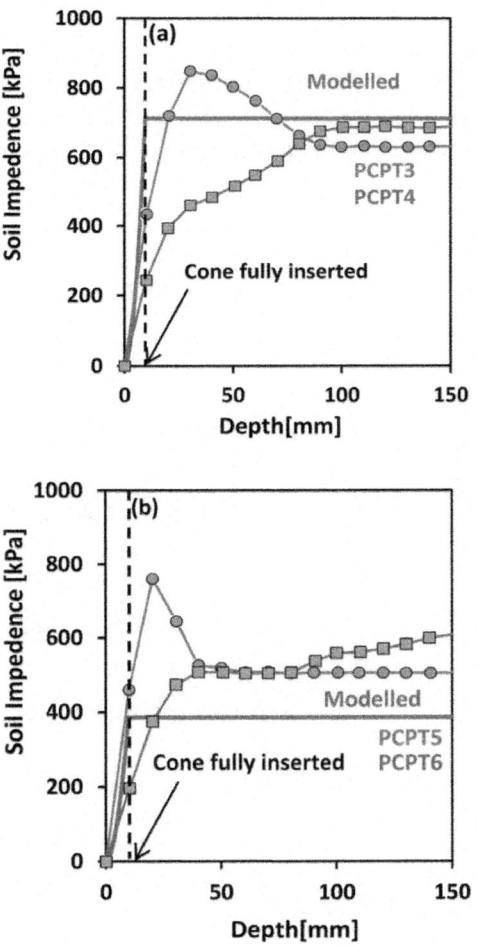

Figure 8. Soil mechanical impedance to cone penetration for different soil mechanical properties and cone types. Two replicates of a silty clay were measured to have (a) $s_u = 65$kPa and $G = 567 \times s_u$; (b) $s_u = 40$kPa and $G = 150 \times s_u$. Experimental data correspond to two tests conducted with duplicate cones of the same geometry but subtle physical

design differences [49]: miniature piezocone penetrometer (PCPT4 and PCPT6) and miniature quasi-static cone penetrometer (PCPT3 and PCPT5). Data points were obtained from [49] with the dashed lines denoting the positions when the cone was fully inserted.

Simulated Forces and Stresses During Penetration-Expansion in Soil

The internal earthworm (or root) pressures required for cavity expansion or for penetration of new soil volumes were computed for various cavity radii, apex insertion angles and combinations of hydration and soil properties listed in Table 2. The results in Fig 9 illustrate that the pressure required for cavity expansion increases for drier soil and reaches the intrinsic maximum pressure (that an earthworm can exert) at normalized water content of roughly 0.1 and 0.2 for soils with clay contents of 16% and 50% respectively.

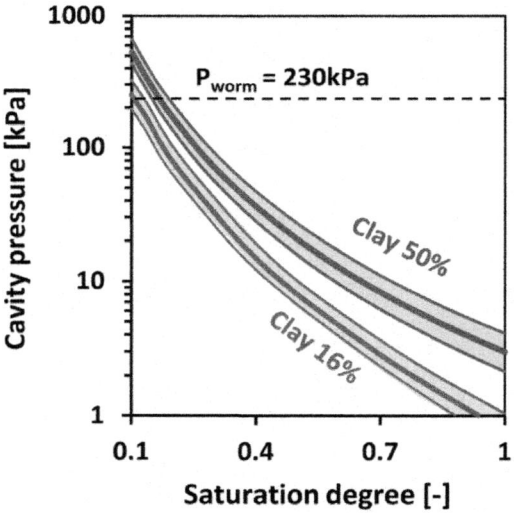

Figure 9. Predicted cylindrical cavity pressure as a function of saturation for two clay contents (16 and 50%) and for a range of cavity radii (1 to 5 mm). The blue and red curves denote soils with clay contents of 16% and 50% respectively, with the hydromechanical correlations presented in Table 3. Thick red and blue lines refer to earthworm radius of 2.5 mm, while the enveloping curves represent radii between 1 and 5 mm.

The exact "cone" geometry has only a small influence on the penetration and cavity expansion pressures. Fig 9b illustrates the maximum penetration resistance pressures for varying semi apex angles for the same moisture contents. The simulations were conducted assuming a frictionless (m = 0) interface between the penetrating object (earthworm or plant root) and the soil. Soil drying (reduced water content) increases cavity limiting pressure and with it the penetration resistance. The results in Fig 10 were computed for soils with a clay content of 16%, but similar trends were obtained for soils with 50% clay content (slightly higher mechanical resistance values). In summary, the penetration pressures increase with increasing clay content, decrease with increasing water content, decrease (slightly) with increasing radius, and slightly increase with increasing semi-apex angle.

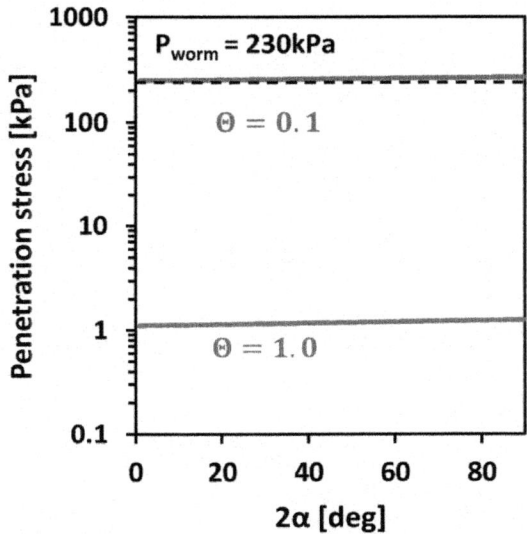

Figure 10. Maximum penetration resistance stress vs. cone apex angles for different normalized water contents with a base radius of 2.5 mm at 16% clay content. Simulations were conducted for normalized water contents of 0.1, and 1.0 at a soil clay content of 16%. Soils with larger clay content display similar mechanical behavior at larger normalized water contents.

Cone penetration resistance in terms of stresses and penetration resistive forces for different earthworm and plant root radii are depicted in Fig 11. Fig 11 (a) shows that the penetration stress decreases (slightly) with increasing radius. In contrast, the penetration force increases with cone radius (Fig 11 (b). The interplay between penetration (resistive) force and stress will be elaborated in the discussion section, in the context of estimating energy costs of bioturbation.

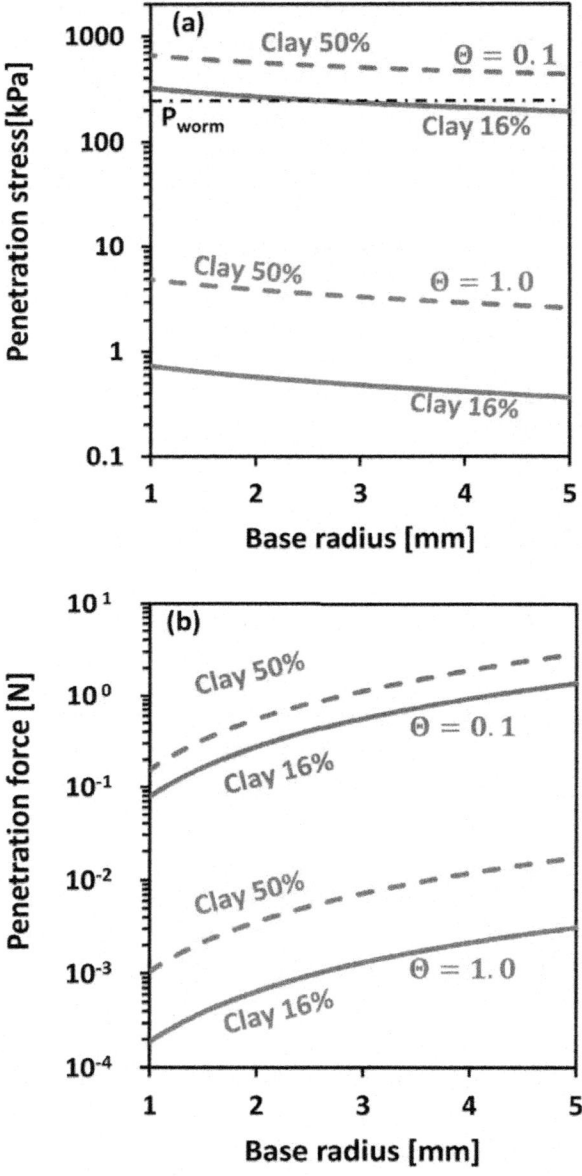

Figure 11. Penetration stress and resistances as a function of base radius. (a) Penetration stress; (b) penetration resistance. Dashed curves denote soils with clay contents of 50%, while solid curves denote clay contents of 16%. Penetration stress (a) decreases for increasing base radius for fixed soil mechanical properties. In contrast, for the same penetration stresses, penetration resistance (measured as axial force) increases with increasing base radius.

Strain Energy of Cavity Expansion in Plastic Soil—Ecological Considerations

Estimates of bioturbation strain energy were computed for different clay contents and normalized water contents, and for radii in the range of 1 to 5 mm, as listed in Table 2. The strain energy density was estimated from strain energy values associated with the minimum earthworm radius of 1 mm. Fig 12 depicts the change in strain energy density as a function of clay content and water content. Applying a conversion coefficient of 0.0484 g_{carbon} J^{-1} to translate energy requirement to soil organic carbon requirement [7], we estimate minimum soil organic carbon (SOC) contents required to support penetration by earthworms in soils with different clay and water contents (Fig 12). The strain energy (and required SOC) decreases with increasing water content, and increases with increasing clay content.

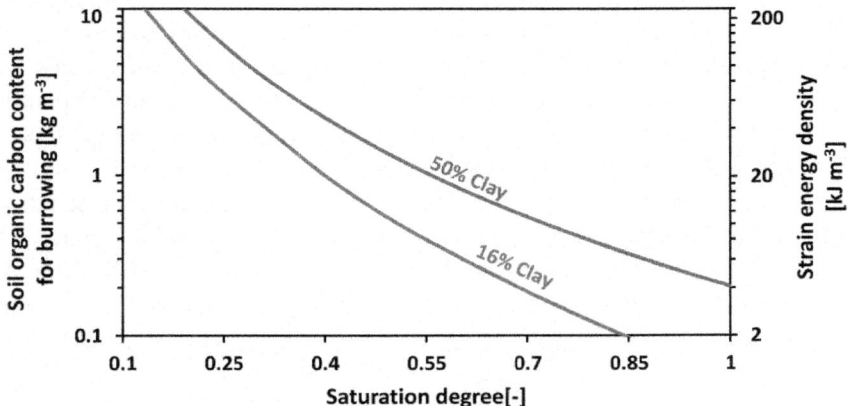

Figure 12. Required soil organic carbon content and maximum strain energy density as a function of normalized water content for clay contents of 16% and 50%. Analysis is based on the steady state mechanical model to determine strain energy. For given soil organic carbon content, one can determine the range of normalized soil moistures under which the energetic demands of displacing a volume of soil are less than the energy stored in the soil organic carbon in the same volume (range to the right of the respective line in the figure). This neglects any rate dependent effects. Note that mechanical energy is mapped to soil organic carbon content using a factor of 1.2/24.8 kg kJ^{-1} [7], and the tick marks on the left vertical axis were spaced in order to align with those on the right vertical axis.

To estimate the annual carbon consumption of earthworm communities, we use published examples of earthworm abundances in different ecosystems (Table 3). For a soil with 16% clay content and normalized water content of 0.4 (representing field capacity [53]), the strain energy density was calculated to be 24.1 kJ m^{-3}. For an average earthworm of 2.5 mm radius, the energy required to create a 1 m long tunnel would amount to 24.1 kJ m^{-3} × π × (2.5 × 10^{-3} m)^2 × 1 m = 4.7 × 10^{-4} kJ. Assuming a mean penetration rate of 0.1 m day^{-1}(or 36.5 m year^{-1}) (Table 3), we estimate an annual mechanical work of 17.2 J year^{-1} per earthworm, which is equivalent to the energy contained in 0.8g of SOC. Assuming a typical earthworm population density in a paddock of 300 individuals per m^3 soil in the top 0.2 m of soil (Table 3) and the same penetration rate of 0.1 m day^{-1} for each individual, we estimate a minimum consumption of 8 × 10^{-4} kg year^{-1} × 300 ind m^{-3} × 0.2 m = 0.05 kg year^{-1} soil organic carbon per m^2 surface area to cover the energy expenditure for soil penetration by the whole earthworm community.

Comparison to Crack Propagation Model

A comparison of energy requirements for creating a 1 m long cavity with a radius of 1.2 mm by plastic penetration (our model) or fracture propagation (model by Dorgan et al. [50]) is presented in Fig 13. For a normalized water content of 1 [-], the crack propagation model predicts an energy expenditure of 3.2 × 10^{-3} J, while the elasto-plastic cavity expansion model presented in this study yields an energy expenditure of 3.1 × 10^{-3} J. For normalized water contents in the range of 0.2 to 1 [-], the strain energy required for penetration based on the elasto-plastic penetration-cavity expansion model is much lower than for fracturing. For normalized water contents below 0.02 [-], the strain energy required for plastic penetration exceeds that of the crack propagation.

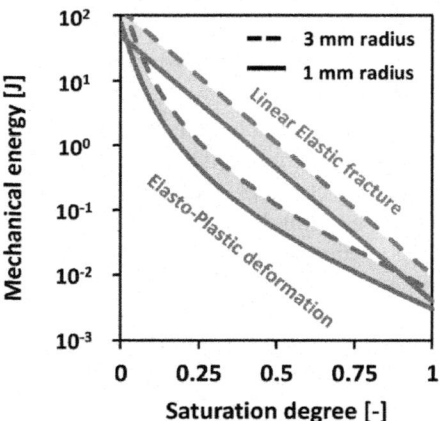

Figure 13. Mechanical energy to create a burrow of 1 m length and 1.2 mm radius as a function of normalized water content using a penetration model and a fracture model. Both models were conducted for a worm of r = 1.2 mm [50] for soils with clay contents ranging from 15–25% [21, 50–52].

DISCUSSION

The theoretical estimates and their general agreement with limited experimental data inspire confidence in using these approximations to estimate the energy requirements for creation of earthworm burrows or plant root channels. The resistance stresses for plant roots and earthworms are reduced with increasing radius (Fig 11 (a)), hence the mechanical energy per unit of displaced soil volume decreases with increasing radius. This implies that larger tunnel radii would be more energy efficient if displacing soil material where the primary goals of earthworms and plant roots. However, if movement through soil in search for resources drives bioturbation, the mechanical energy required to traverse a certain distance in the subsurface may be ecologically more relevant for plant roots and earthworms. The results in Fig 11 (b), depict an increase in penetration force with increasing cavity radii, and the mechanical energy per distance travelled in the soil increases with radius. Hence, while smaller earthworms or plant roots are less energy-efficient in terms of volumetric soil displacement, worms or roots with smaller radii are more energy-efficient for traversing the subsurface.

A decrease in strain energy for wetter soils is attributed to the reduction in soil strength and shear modulus with increasing water content. The strain energy requirements increase with increasing soil clay contents that affect soil strength and shear modulus. There is a slight increase in penetration resistance with increasing semi-apex insertion angle. The axial component of forces acting normal to the face of an earthworm or plant root increases with increased apex angles resulting in an increase in penetration resistance.

The plastic earthworm penetration model presented in this study compared well with the linear elastic fracture mechanics (LEFM) model of Dorgan et al. [50] at high water contents, predicting a mechanical energy expenditure of 3.1×10^{-3} J, whereas Dorgan et al. [50] estimates 3.2×10^{-3} J for an earthworm to propagate a crack of the same length. Despite many differences in the modeling approaches and underlying assumptions, the mechanical energy computed for crack propagation and penetration-expansion are similar for water contents near saturation [50].

For intermediate water contents, the LEFM model typically results in a larger expenditure per unit of cavity length than the plastic penetration model, while for very low saturation degrees (below 0.02 [-]), fracture propagation appears to be less energy consuming than plastic penetration (Fig 13). However, LEFM would imply that creation of permanent fractures at saturation degrees below 0.4 [-] would require cavity pressures in excess of 230 kPa, which is the physiological limit of earthworms. The results suggest that plastically deforming the soil is probably more energy efficient for wet soils, while for dry soils (saturation degrees lower than 0.02 [-]), Fig 13 suggests that crack propagation is energetically advantageous over penetration-cavity expansion. These preliminary results suggest that different modes of soil penetration may be favorable for different soil water contents (or soil mechanical properties).

Plant roots have been observed to expand radially upon encountering soils with high mechanical impedance, and in this way weakening the forefront [15], potentially inducing a brittle failure close to the root tip, allowing for axial elongation into an open crack. For the case of this soil, the water contents where linear elastic fracture mechanics seems more energetically feasible than elasto-plastic deformation is below the physiological limit where plant roots can actually uptake water (corresponding to a saturation degree of 0.2 [-]), that is the matric potential of the soil exceeds the osmotic pressures that plant roots can extract water from the root soil interface (root suction pressures reported to range from −2.5 to −3.5 MPa [54]). It has been shown that soil plastic deformation and crack propagation are not mutually exclusive phenomena (Yoshida and Hallett [55]), and we note that more systematic studies are required to delineate respective envelopes of applicability of fracture mechanics and plastic deformation models. Despite neglecting potential crack formation, the elastic-perfectly plastic model presented in this study still determines a minimum energy requirement necessary for bioturbation.

The model developed in the present study could be used to assess the amount of soil organic carbon (SOC) necessary to support earthworm activity in soil. Fig 12 depicts estimates for the minimum amount of SOC required for an individual earthworm to satisfy the energy demand of soil penetration (for clay contents of 16% and 50%). Based on these model estimates, earthworm penetration in soil with normalized water contents below 0.25 can only be economical if SOC content exceeds 10 kg m⁻³. This value seems realistic, as Crumsey et al. [56] also found 10 kg m⁻³ SOC in their mesocosm experiments, which were maintained at field capacity during the experiments. In order to provide conservative estimations of energetic costs of cavity creation, our analysis presented in the results section was based on normalized water content at field capacity ($\Theta = 0.4$).

Fig 12 implies that the energetic costs for constructing a 1 m long tunnel by an earthworm with 2.5 mm radius [3, 42] at a normalized water content of 0.4 and 16% clay content are equivalent to the energy contained in roughly 2.3 $\times$ 10⁻⁵ kg soil organic carbon. Considering an average earthworm maintenance respiration rate at rest of 3.6 $\times$ 10⁻⁶ kg day⁻¹ (of carbon) [57], this implies that the strain energy required to burrow 1 m distance is equivalent to 7 days of maintenance respiration at rest. Provided the upper limit of earthworm penetration rate is 0.2 m day⁻¹ (Table 3) this would suggest that soil penetration could account for up to half of an earthworm's energy budget.

Scaling up of our energy estimates to the burrowing activity of a typical earthworm community in a paddock resulted in energy expenditures equivalent to as much as 0.05 kg year^{-1} SOC per m^2 surface area. Alban and Berry [58] witnessed invasive earthworms depleting SOC at a rate of 0.06 kg year^{-1} per m^2, suggesting that our estimates are in close agreement with field observations. Considering that, Beer et al. [59] estimated the net primary productivity (NPP) in cropland to be in the order of 0.9 kg m^{-2}, the estimates above suggest that earthworm populations could consume as much as 5% of the annual NPP in crop lands. Taking the lowest normalized water content for which soil penetration would still be economically feasible with an SOC concentration of 10 kg m^{-3} (Θ = 0.25 in Fig 12), a similar calculation would yield that the earthworm community energy expenditure for the same burrowing activity would be equivalent to as much as 0.15 kg year^{-1} SOC per m^2, which would be equivalent to 15% of the annual NPP in crop lands. However, note that these estimates are based on constant soil water content, whereas natural hydration dynamics would affect the energetic costs of soil penetration as well as bioturbation activity and earthworm population sizes.

An important aspect not considered in this analysis is the role of soil ingestion by earthworms, an essential process for extracting the required soil organic carbon and contributing to many soil regulatory functions [60]. Ingested soil passing through the earthworm gut stimulates microbial activity, enhances aggregation by mucus secretion and litter collection, and gives rise to mutual interactions between microbes and earthworms. On the other hand, earthworms ingest a wide variety of organic matter including fungus and bacteria [60, 61]. Evidence suggests beneficial impact on plant root activities (despite anecdotal evidence for earthworms ingesting living roots [61]). The partitioning between ingestion and soil displacement is not well understood. Future studies will endeavour to elucidate the mechanical ramifications of ingestion, and, equally important, the role of rapid penetration by earthworms in comparison to slow growth by plant roots that would incur different energetic costs (for the same burrow geometry) due to effects of visco-elastic displacement processes [21] not addressed in this study.

Finally, the results presented in this study suggest that biotic soil processes perform considerable mechanical work on soil, implying that they are likely effective contributors to soil structural restoration after disturbance, e.g. by soil compaction. Abiotic processes such as wetting and drying induce shrinking and swelling [62], which also contributes to expanding cavities. Spherical soil aggregates undergo coalescence due to wetting at axial strain rates of 0.005 m m^{-1} s^{-1}, which is equivalent to the rate of spherical radial expansion due to aggregate swelling [62]. Plant roots expand cavities at strain rates up to 0.08 m m^{-1} s^{-1} [23]. This would imply that macropore creation by biotic processes occurs at rates nearly 20 times faster than that by abiotic processes, thus earthworms and plant roots have the capacity to accelerate soil structural restoration by at least an order of magnitude compared to abiotic processes only. More experimental and theoretical work is needed to understand the interplay between abiotic and biotic processes during post-compaction soil structural restoration. In particular, investigation of the initiation of the process, i.e. the creation of the first channels or cracks, and their benefit for subsequent colonization by roots and earthworms is important, in order to enable quantitative prediction of the speed of recovery of compacted soils.

ACKNOWLEDGMENTS

We acknowledge the collaboration with Thomas Keller (Agroscope) and Achim Walter (ETH Zurich). We would also like to acknowledge Paul Hallett (University of Aberdeen), who's helpful comments and suggestions have tremendously improved the manuscript.

AUTHOR CONTRIBUTIONS

Analyzed the data: SR DO SJS. Contributed reagents/materials/analysis tools: SR DO SJS. Wrote the paper: SR DO SJS. Main idea: DO. Theoretical development: SR SJS. Mathematical Model implementation: SR.

REFERENCES

1.   Materechera S, Alston A, Kirby J, Dexter A. Influence of root diameter on the penetration of seminal roots into a compacted subsoil. Plant and Soil. 1992;144(2):297–303. doi: 10.1007/BF00012888.

2.   Capowiez Y, Cadoux S, Bouchand P, Roger-Estrade J, Richard G, Boizard H. Experimental evidence for the role of earthworms in compacted soil regeneration based on field observations and results from a semi-field experiment. Soil Biology and Biochemistry. 2009;41(4):711–717. doi: 10.1016/j.soilbio.2009.01.006.

3.   Bottinelli N, Jouquet P, Capowiez Y, Podwojewski P, Grimaldi M, Peng X. Why is the influence of soil macrofauna on soil structure only considered by soil ecologists? Soil and Tillage Research. 2014. doi: 10.1016/j.still.2014.01.007

4.   Greacen EL, Sands R. Compaction of forest soils. A review. Soil Research. 1980;18(2):163–189. doi: 10.1071/SR9800163.

5.   Kroener E, Zarebanadkouki M, Kaestner A, Carminati A. Nonequilibrium water dynamics in the rhizosphere: How mucilage affects water flow in soils. Water Resources Research. 2014. doi: 10.1002/2013WR014756.

6.   Gray J, Lissmann H. An apparatus for measuring the propulsive forces of the locomotory muscles of the earthworm and other animals. Journal of Experimental Biology. 1938;15(4):518–521.

7.   Lavelle P, Barot S, Blouin M, Deca"ens T, Jimenez JJ, Jouquet P. Earthworms as key actors in self-organized soil systems. Theoretical Ecology Series. 2007;4:77–106. doi: 10.1016/S1875-306X(07)80007-4.

8.   Capowiez Y, Belzunces L. Dynamic study of the burrowing behaviour of Aporrectodea nocturna and Allolobophora chlorotica: interactions between earthworms and spatial avoidance of burrows. Biology and Fertility of Soils. 2001;33(4):310–316. doi: 10.1007/s003740000327.

9.   Nadelhoffer KJ, Aber JD, Melillo JM. Fine roots, net primary production, and soil nitrogen availability: a new hypothesis. Ecology. 1985;66(4):1377–1390. doi: 10.2307/1939190.

10. Greacen E, Oh J. Physics of root growth. Nature. 1972;235(53):24–25. doi: 10.1038/newbio235024a0

11. Atwell B. Response of roots to mechanical impedance. Environmental and Experimental Botany. 1993;33(1):27–40. doi: 10.1016/0098-8472(93)90053-I.

12. Misra R, Dexter A, Alston A. Maximum axial and radial growth pressures of plant roots. Plant and Soil. 1986;95(3):315–326. doi: 10.1007/BF02374612.

13. Bengough A, Mullins C. Mechanical impedance to root growth: a review of experimental techniques and root growth responses. Journal of Soil Science. 1990;41(3):341–358. doi: 10.1111/j.1365-2389.1990.tb00070.x.

14. Dexter A. Mechanics of root growth. Plant and Soil. 1987;98(3):303–312. doi: 10.1007/BF02378351.

15. Hettiaratchi D, Goss M, Harris J, Nye P, Smith K. Soil Compaction and Plant Root Growth [and Discussion]. Philosophical Transactions of the Royal Society of London Series B: Biological Sciences. 1990;329(1255):343–355. doi: 10.1098/rstb.1990.0175.

16. Gabet EJ, Mudd SM. Bedrock erosion by root fracture and tree throw: A coupled biogeomorphic model to explore the humped soil production function and the persistence of hillslope soils. Journal of Geophysical Research: Earth Surface (2003–2012). 2010. doi: 10.1029/2009jf001526

17. McKenzie BM, Dexter AR. Radial pressures generated by the earthworm Aporrectodea rosea. Biology and Fertility of Soils. 1988 Jan;5(4):328–332. Available from: http://link.springer.com/article/10.1007/BF00262141. doi: 10.1007/BF00262141.

18. Dexter A. Model experiments on the behaviour of roots at the interface between a tilled seed-bed and a compacted sub-soil. Plant and Soil. 1986;95(1):149–161. doi: 10.1007/BF02378860.

19. Dorgan KM, Arwade SR, Jumars PA. Burrowing in marine muds by crack propagation: kinematics and forces. Journal of Experimental Biology. 2007;210(23):4198–4212. doi: 10.1242/jeb.010371. pmid:18025018

20. Murphy EA, Dorgan KM. Burrow extension with a proboscis: mechanics of burrowing by the glycerid Hemipodus simplex. The Journal of Experimental Biology. 2011;214(6):1017–1027. doi: 10.1242/jeb.051227. pmid:21346130

21. Ghezzehei TA, Or D. Rheological properties of wet soils and clays under steady and oscillatory stresses. Soil Science Society of America Journal. 2001;65(3):624–637. doi: 10.2136/sssaj2001.653624x.

22. Quillin K. Ontogenetic scaling of burrowing forces in the earthworm Lumbricus terrestris. Journal of Experimental Biology. 2000;203(18):2757–2770. pmid:10952876

23. Dumais J, Forterre Y. Vegetable dynamicks: the role of water in plant movements. Annual Review of Fluid Mechanics. 2012;44:453–478. doi: 10.1146/annurev-fluid-120710-101200.

24. Pritchard J. The control of cell expansion in roots. New Phytologist. 1994;127(1):3–26. doi: 10.1111/j.1469-8137.1994.tb04255.x.

25. Gerard C, Sexton P, Shaw G. Physical factors influencing soil strength and root growth. Agronomy journal. 1982;74(5):875–879. doi: 10.2134/agronj1982.00021962007400050025x.

26. Wiersum L. The relationship of the size and structural rigidity of pores to their penetration by roots. Plant and Soil. 1957;9(1):75–85. doi: 10.1007/BF01343483.

27. Scholefield D, Hall D. Constricted growth of grass roots through rigid pores. Plant and Soil. 1985;85(2):153–162. doi: 10.1007/BF02139621.

28. Keudel M, Schrader S. Axial and radial pressure exerted by earthworms of different ecological groups. Biology and Fertility of Soils. 1999;29(3):262–269. doi: 10.1007/s003740050551.

29. Bishop R, Hill R, Mott N. The theory of indentation and hardness tests. Proceedings of the Physical Society. 1945;57(3):147. doi: 10.1088/0959-5309/57/3/301.

30. Carter JP, Booker JR, Yeung SK. Cavity expansion in cohesive frictional soils. G'eotechnique. 1986 Jan;36(3):349–358. Available from:http://www.icevirtuallibrary.com/content/article/10.1680/geot.1986.36.3.349.

31. Yu H. Discussion:Singular Plastic Fields in Steady Penetration of a Rigid Cone(Durban, D., and Flek, NA, 1992, ASME J. Appl Mech., 59, pp. 706–710). Journal of Applied Mechanics. 1993;60(4):1061–1062. doi: 10.1115/1.2900981.

32. Durban D, Fleck NA. Singular plastic fields in steady penetration of a rigid cone. Journal of Applied Mechanics. 1992;59(4):706–710. doi: 10.1115/1.2894032.

33. Shames IH. Elastic and inelastic stress analysis. CRC Press; 1997.

34. Elder DM. Stress Strain and Strength Behaviour of Very Soft Soil Sediment. University of Oxford; 1985.

35. Kemper W, Rosenau R. Soil cohesion as affected by time and water content. Soil Science Society of America Journal. 1984;48(5):1001–1006. doi: 10.2136/sssaj1984.03615995004800050009x.

36. Fredlund D, Morgenstern N, Widger R. The shear strength of unsaturated soils. Canadian Geotechnical Journal. 1978;15(3):313–321. doi: 10.1139/t78-029.

37. Bowles JE. Foundation analysis and design. McGraw-Hill, London; 1988.

38. Abdalla A, Hettiaratchi D, Reece A. The mechanics of root growth in granular media. Journal of Agricultural Engineering Research. 1969;14(3):236–248. doi: 10.1016/0021-8634(69)90126-7.

39. Cokca E, Erol O, Armangil F. Effects of compaction moisture content on the shear strength of an unsaturated clay. Geotechnical & Geological Engineering. 2004;22(2):285–297. doi: 10.1023/B:GEGE.0000018349.40866.3e.

40. Newell G. The role of the coelomic fluid in the movements of earthworms. Journal of Experimental Biology. 1950;27(1):110–122.

41. VandenBygaart A, Fox C, Fallow D, Protz R. Estimating earthworm-influenced soil structure by morphometric image analysis. Soil Science Society of America Journal. 2000;64(3):982–988. doi: 10.2136/sssaj2000.643982x.

42. Ehlers W. Observations on earthworm channels and infiltration on tilled and untilled loess soil. Soil Science. 1975;119(3):242–249. doi: 10.1097/00010694-197503000-00010.

43. Daniel O. Population dynamics of lumbricus terrestris L.(oligochaeta: lumbricidae) in a meadow. Soil Biology and Biochemistry. 1992;24(12):1425–1431. doi: 10.1016/0038-0717(92)90128-K.

44. Chan K. An overview of some tillage impacts on earthworm population abundance and diversityimplications for functioning in soils. Soil and Tillage Research. 2001;57(4):179–191. doi: 10.1016/S0167-1987(00)00173-2.

45. Fragoso C, Lavelle P. Earthworm communities of tropical rain forests. Soil Biology and Biochemistry. 1992;24(12):1397–1408. doi: 10.1016/0038-0717(92)90124-G.

46. Or D, Wraith JM. Temperature effects on soil bulk dielectric permittivity measured by time domain reflectometry: A physical model. Water Resources Research. 1999;35(2):371–383. doi: 10.1029/1998WR900008.

47. COMSOL I. Comsol Reference Manual. New England Executive Park Burlington, MA 01803 USA: Coms Inc.; 2012.

48. Walker J, Yu H. Adaptive finite element analysis of cone penetration in clay. Acta Geotechnica. 2006;1(1):43–57. doi: 10.1007/s11440-006-0005-9.

49. Kurup P, Voyiadjis G, Tumay M. Calibration chamber studies of piezocone test in cohesive soils. Journal of Geotechnical Engineering. 1994;120(1):81–107. doi: 10.1061/(ASCE)0733-9410(1994)120:1(81).

50. Dorgan KM, Lefebvre S, Stillman JH, Koehl M. Energetics of burrowing by the cirratulid polychaete Cirriformia moorei. Journal of Experimental Biology. 2011;214(13):2202–2214. doi: 10.1242/jeb.054700. pmid:21653814

51. Hanson JA, Hardin BO, Mahboub K. Fracture toughness of compacted cohesive soils using ring test. Journal of geotechnical engineering. 1994;120(5):872–891. doi: 10.1061/(ASCE)0733-9410(1994)120:5(872).

52. Wang JJ, Zhu JG, Chiu C, Zhang H. Experimental study on fracture toughness and tensile strength of a clay. Engineering Geology. 2007;94(1):65–75. doi: 10.1016/j.enggeo.2007.06.005.

53. Cooper GS, Smith R. Sequence of products formed during denitrification in some diverse western soils. Soil Science Society of America Journal. 1963;27(6):659–662. doi: 10.2136/sssaj1963.03615995002700060027x.

54. Lambers H, Chapin FS III, Pons TL. Plant water relations. Springer; 2008.

55. Yoshida S, Hallett P. Impact of hydraulic suction history on crack growth mechanics in soil. Water Resources Research. 2008;44(5). doi: 10.1029/2007WR006055.

56. Crumsey JM, Le Moine JM, Capowiez Y, Goodsitt MM, Larson SC, Kling GW, et al. Community-specific impacts of exotic earthworm invasions on soil carbon dynamics in a sandy temperate forest. Ecology. 2013;94(12):2827–2837. doi: 10.1890/12-1555.1. pmid:24597228

57. Speratti AB, Whalen JK. Carbon dioxide and nitrous oxide fluxes from soil as influenced by anecic and endogeic earthworms. Applied Soil Ecology. 2008;38(1):27–33. doi: 10.1016/j.apsoil.2007.08.009.

58. Alban DH, Berry EC. Effects of earthworm invasion on morphology, carbon, and nitrogen of a forest soil. Applied Soil Ecology. 1994;1(3):243–249. doi: 10.1016/0929-1393(94)90015-9.

59. Beer C, Reichstein M, Ciais P, Farquhar G, Papale D. Mean annual GPP of Europe derived from its water balance. Geophysical Research Letters. 2007;34(5). doi: 10.1029/2006GL029006.

60. Brown GG, Barois I, Lavelle P. Regulation of soil organic matter dynamics and microbial activity in the drilosphere and the role of interactions with other edaphic functional domains. European Journal of Soil Biology. 2000;36(3):177–198. doi: 10.1016/S1164-5563(00)01062-1.

61. Curry JP, Schmidt O. The feeding ecology of earthworms–a review. Pedobiologia. 2007;50(6):463–477. doi: 10.1016/j.pedobi.2006.09.001.

62. Ghezzehei TA, Or D. Dynamics of soil aggregate coalescence governed by capillary and rheological processes. Water Resources Research. 2000;36(2):367–379. doi: 10.1029/1999WR900316.

Chapter 11

RESPONSE CHARACTERISTICS OF SOIL FRACTAL FEATURES TO DIFFERENT LAND USES IN TYPICAL PURPLE SOIL WATERSHED

Bang-lin Luo[1], Xiao-yan Chen[1], Lin-qiao Ding[1], Yu-han Huang[1], Ji Zhou[2], and Tian-tian Yang[3]

[1]College of Resources and Environment/Key Laboratory of Eco-environment in Three Gorges Region (Ministry of Education), Southwest University, Chongqing, China

[2]State Key Lab of Urban and Regional Ecology, Research Centre for Eco-Environmental Sciences, Chinese Academy of Sciences, Beijing, China

[3]Department of Civil and Environmental Engineering, University of California Irvine, Irvine, California, United States of America

ABSTRACT

As a fundamental characteristic of soil physical properties, the soil Particle Size Distribution (PSD) is important in the research on soil moisture migration, solution transformation, and soil erosion. In this research, the PSD characteristics with distinct methods in different land uses are analyzed. The results show that the upper bound of the volume domain of the clay domain ranges from 5.743 μm to 5.749 μm for all land-use types. For the silt domain of purple soil, the value ranges among 286.852~286.966 μm. For all purple soil land-use types, the order of the volume domain fractal dimensions is $D_{clay} < D_{silt} < D_{sand}$. However, the values of D_{silt} and D_{sand} in the *Pinus massoniana Lamb, Robinia pseudoacacia L* and*Ipomoea batatas* are all higher than the corresponding values in the *Citrus reticulate Blanco*and *Setaria viridis*. Moreover, in all the land-use types, all of the parameters in volume domain fractal dimension (D_{vi}) are higher than the corresponding parameter values from the United States Department of Agriculture ($D_{vi}(U)$). The correlation study between the volume domain fractal dimension and the soil properties shows that the intensity of correlation to the soil texture and soil organic matter has the order as: $D_{silt} > D_{silt}(U) > D_{sand}(U) > D_{sand}$ and $D_{silt} > D_{silt}(U) > D_{sand}(U) > D_{sand}(U)$, respectively. As it is compared with all D_{vi}, the D_{silt} has the most significant correlativity to the soil texture and organic matter in different land uses of the typical purple soil watersheds. Therefore, D_{silt} will be a potential indictor for evaluating the proportion of fine particles in the PSD, as well as a key measurement in soil quality and productivity studies.

INTRODUCTION

The fractal theory was proposed and established by Mandelbrot (1977, 1982) [1–2], which is a method of describing systems with non-characteristic scales and self-similarity. This theory has been utilized to quantitatively describe the characteristics of the soil particle size distribution (PSD), which is important in hydrological conductivity, solution transportation, and soil erosion. Therefore, the fractal theory attracts the interests of pedologists worldwide [3–7]. In the area of the micro-field of soil science, Scott and Stephen estimated soil water retention based on fractal mathematics and further noted the limitations of the fractal method when applied to scaling the soil PSD [8–9]. The method of measuring the soil PSD was developed with the introduction of lighting-scattering technology, which was utilized to further quantitatively calculate the soil particle size distribution [10]. Almost at the same time, with the enhancement of the PSD measurement technology, the fractal theory was also developed into multifractal theory. Perfect and Kay (1993) [11] analyzed the characteristics of soil aggregate fragmentation and provided a theoretical framework in terms of the combination of fractals and multifractals to link size distribution and strength. Bittelli et al. (1999) [12] characterized the PSD using a fragmentation model based on the mass fractal dimension (D_m) in three domains: clay, silt, and sand. Later, Prosperinin (2008) [13] systematically described the characteristics of PSD in the Umbria region of Italy using the fractal mathematics method. Unlike traditional factors used to determine the characteristics of hydraulic properties of soil, Eran (2009) and Lalit (2010) applied the fractal theory to predict the soil hydraulic properties through the measurement of PSD. Some scholars also applied the multifractal model on PSD. Posadas et al. (2001, 2003, 2010) [14–16] characterized the soil particle distribution by applying the multifractal method, interpreted the relationship between the soil textural features and parameters of the multifractal model, and described the soil pore system based on the multifractal method. The studies from Posadas et al. (2001, 2003, 2010) [14–16] have improved the integration of spatial properties of the soil pore system. In addition, Miguel (2002) applied the multifractal model to characterize the soil volume-size distribution. Caniego (2005) [17] also focused on the soil spatial properties in terms of organic matter and electrical conductivity using multifractal theory.

The fractal and multifractal models became even more important for quantitatively describing the characteristics and behaviors of the soil particles with the development of laser diffraction technology and the stochastic methods based on probability theories, which is more reasonable in describing the features of the soil solid phase [18]. These developments provide a necessary environment for the movement of soil liquid and gas phases, all of which strongly affect the eco-hydrological processes. Soil fractal dimension is an important parameter to study soil structure, soil texture and soil erosion. It has an important theoretical value and practical significance to study the soil fractal dimension in the purple soil distributed region.

In China, Yang et al. (1993) [19] introduced a calculation theory concerning the soil PSD mass fractal dimension (D_m). Huang (2002, 2005) [20–21] analyzed the relationship between the soil particle size mass distribution and the soil compartments, including the clay, silt, and sand contents. The results from Huang (2002, 2005) [20–21] indicated that the mass-based fractal dimension of the soil PSD could be used to predict the soil water retention properties. Huang further fitted the relationship between the fractal dimension and soil texture and evaluated the fractal distribution with a water retention curve. However, Martin and Montero (2002) [22] questioned the assumption that soil particles with different sizes have the same density in the D_m calculations. In order to avoid this issue, another approach to calculate the soil particle fractal dimension and the volume fractal dimension D_v, was proposed by Wang et al. (2005) [23]. Since then, fractal theory has become well-developed, as pointed out by Yang (2008) [24], in which the comparison and analysis of the mass fractal dimension D_m and the volume fractal dimension D_v were presented. The fractal and multifractal theory have been also widely applied to various topics in soil science, including land use, desertification, and the characteristics of specific soil particles [7,24–27]. Wang (2008) [28] applied the multifractal theory to analyze the effect of some parameters in the multifractal model on the different land-use types in the Loess Plateau of China, and Liu (2009) [29] focused on the effect of the fractal features of soil PSD on different plant communities in Chinese forests. With regards to the studies mentioned above, very few studies have been conducted on the response of the PSD fractal features to the different land use types in the small purple soil watershed in southwestern China. Zhang (2008) [30] analyzed the soil aggregate distribution through the fractal method to characterize the purple soil of the Sichuan Basin and further discussed the impact of vegetation type on soil particles' water stability, soil aggregates' mechanics, and soil chemical stability in the Sichuan Basin. However, the volume domain fractal dimension D_v are obtained by laser diffraction, and the volume domain fractal dimension features of different land use types in the small watershed of purple soil have never been reported. As one of the most erodible and productive soil types in China, purple soil most distributed in southwestern China. Therefore, there is great significance to understand the response characteristics of soil fractal features.

As a new method to quantitatively describe the soil structure, the dimensionless soil fractal dimension is able to be more easily and efficiently estimated. The goal is to construct the relationship between soil PSD and the soil properties, such as soil texture, soil aggregate stability, and hydraulic conductivity. These properties will greatly affect the soil particle characteristics based on the ecological and hydrological processes in different land uses. Moreover, due to the fact that the fractal dimension is based on the self-similarity theory, which is regarded as useful way to describe the soil particle and the soil pore system [8], the fractal dimension theory is a reasonable method to evaluate the soil particle and relevant characteristics. This paper analyzes the measurement of PSD characteristics in different land uses of typical purple soil watersheds in terms of the calculation of the volume fractal dimension D_v, including D_{clay}, D_{silt}, and D_{sand}. The objective of this paper is to correlate the soil textures, soil organic matter, and PSD volume fractal dimension, providing the indicator for better evaluating texture, quality, and productivity of purple soil in different land uses.

MATERIALS AND METHODS

Study Area Description

The study is carried out on private land of each location (Fig 1), with the permissions from the land owners. All of the land uses are either for tillage or for landscaping with economic trees, and no specific permissions were required. According to the field investigation, the field does not involve with endangered or protected species. (Fig 1, a picture shows the research areas.)

Two typical small purple soil watersheds, Yangjiagou (YJG) and Daijiagou (DJG), are chosen as the research areas. YJG is located between 108°30'18"E and 30°44'36"N, and the DJG is a small watershed located between 108°30'19"E and 30°44'21"N. The contour map in Fig 1 shows that the hillslopes of YJG and DJG, which have a south-northerly and east-westerly aspect, respectively. The areas of YJG and DJG are 60.97 hm² and 66.37 hm², respectively, and the altitude ranges from 422.08 m to 811.30 m for both catchments. The geomorphological compositions of the two typical purple soil dominated small watersheds mainly consist of low mountains and low hills, which are common in the Three Gorges reservoir area. Moreover, the purple soil rock types in the two study areas are mainly gray-brown purple sand mudstone of Shaximiao Formation (J_2S). The development of the soil into a gray-brown purple soil indicates a weak erosion resistance capacity. The climates of the two research areas are mainly subtropical moist monsoon. Affected by the southeast and the southwest monsoons, and the two experimental areas have an average temperature of 17°C.

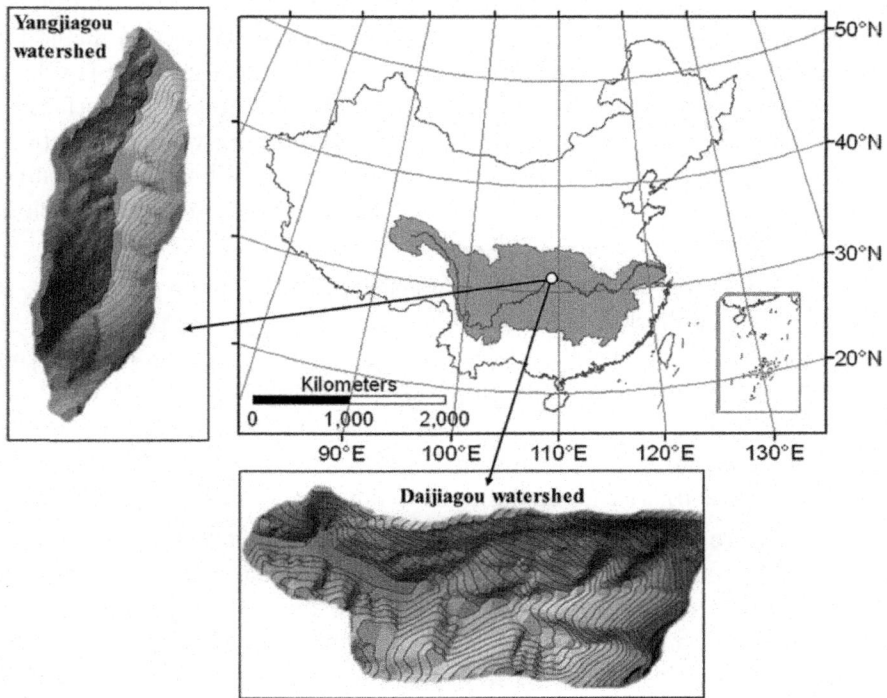

Figure 1. Study areas: the Daijiagou small watershed and the Yangjianggou watershed.

The approximately annual average precipitation is 1000–1350 mm, most of which occurs during April-October. Based on the field investigation, the vegetation in the research areas mainly includes *Pinus massoniana Lamb*, *Robinia pseudoacacia L*, and *Citrus reticulate Blanco*, which represent the forest land use, and a small portion of *Zea mays L* and *Ipomoea batatas, and Setaria viridis*. The *Zea mays L* and *Ipomoea batatas* are the representatives of the farmland types, which are affected by human management activities, and the *Setaria viridis* represents the grassland types.

Soil Sampling and Analysis Methods

Based on the field vegetation and pedologic survey of the land-use types in the YJG and DJG catchments, forestlands (*Pinus massoniana Lamb, Robinia pseudoacacia L*, and *Citrus reticulata Blanco*), farmlands (*Zea mays L* and

Ipomoea batatas) and grassland (*Setaria viridis*) are chosen as the research targets. Multiple surface soil samples are collected from a depth of 20 cm in the six main vegetation areas of both catchments. The detailed soil sample collection method can be found in Zhang et al. (2008) [30–31], in which a combination of the various land-use areas in the research areas with the soil collection method is presented. The major steps of the collection method are described as follows: (1) a specific sample point in a land-use type is selected as a center point; (2) within a 10~15 m radius of this point, 4~6 surface soil samples are collected at randomly chosen locations, allowing the location is random and stochastic; and (3), these ambient point soil samples are mixed to produce an approximately 1 kg soil sample. Totally, there are 178 such soil samples are obtained from the YJG and DJG catchments.

The following analyses, which consist of three main procedures, are carried out using the soil samples.

a.   Dry the samples. According to experiments from Bittlli and Wang (1999, 2007) [12,32], all the soil samples are oven-dried for 24 h at 105°C. And then, the samples are slightly crushed, and the fine remnant roots in the dry soil samples are removed. The oven-dried samples (approximately 0.50 g each) are filtered through a 2 mm sieve.

b.   Measure Soil PSD. The carbonates are removed from the dry soil samples using hydrochloric acid, sodium acetate, and deionized water. Peroxide (30%, w/w) is also added into the soil samples at 75°C for at least 2 h to remove the organic matter from the soil samples. And then, the acid is removed from the soil sample solution with ultra-pure water. The soil sample solution is further adjusted to a neutral pH. The samples are then immersed in [(NaPO$_3$)$_6$] solution for 20 h. Finally, the sample solutions are ultrasonically dispersed for 5 min and the soil PSD is calculated by*Malvern Mastersize 2000* laser diffusion (UK). The volume percentages of the soil particles are repeatedly measured by the *Mastersize 2000* based on the United States Department of Agriculture (USDA) classification of soil particle size. The soil mainly featured 7 soil particle size distributions: ultra-coarse sand (2–1 mm), coarse sand (1–0.5 mm), medium sand (0.5–0.25 mm), fine sand (0.25–0.1 mm), ultra-fine sand (0.1–0.05 mm), silt (0.05–0.002 mm), and clay (<0.002 mm).

c.   Calculate other soil physical and chemical properties. The soil bulk density is measured by the core cutter method (ISS, 1978), and the content of soil organic matter is determined by the potassium dichromate and sulfuric acid (ISS, 1980), which yields a value of 1.724 times that of organic carbon. The soil particle and organic matter (organic carbon) characteristics of the main land use types in the watershed are listed in Table 1.

Table 1. The soil texture and organic matter by land-use type.

Land-use type	Soil texture						Organic matter (g/kg)	
	Clay content		Silt content		Sand content			
	Mean (%)	CV (%)	Mean (%)	CV (%)	Mean (%)	CV (%)	Mean (%)	CV (%)
Pinus massoniana Lamb	3.61a	42.788	37.32c	20.485	59.07a	15.511	14.78a	43.987
Robinia pseudoacacia L	3.83a	25.613	42.97bc	14.518	53.20ab	13.424	13.60a	32.500
Citrus reticulata Blanco	3.48a	38.426	42.49bc	29.579	54.02ab	25.558	9.10b	24.477
Zea mays L	4.86a	26.880	54.11a	19.902	41.03c	29.390	12.07ab	36.702
Ipomoea batatas	3.79a	18.698	50.12ab	10.850	46.09bc	13.275	13.19ab	22.374
Setaria viridis	3.73a	22.854	44.07abc	18.581	52.20abc	17.192	8.73b	19.508

Footnotes: CV, coefficient of variation. Average values were analyzed by DUNCAN multiple comparisons, and different lowercase letters represent significant differences (P<0.05) between each pair.

doi:10.1371/journal.pone.0122842.t001

Fractal Dimension Model for Calculation

1.   Mass fractal dimension (D_m) and volume fractal dimension (D_v).

According to Yang et al. (2008) [24], the soil particle fractal dimension can be calculated in one of two ways: the soil PSD mass fractal dimension D_m [19] and the volume fractal dimension D_v. Due to the assumption that the soil particles with different sizes have of the same density [22], the use of the volume fractal dimension D_v has been gradually adopted by other researchers. The volume-based method avoids the controversy originated from the assumption that the soil particles have self-similarity and fractal characteristics. Therefore, Wang et al. (2005) [23], further noted that D_v was also an intrinsic property of soil particles in analyzing the distribution of D_v in different land uses. Applications of the volume-based method on Yixing, Jiangsu province, China was also presented by Wang et al. (2005) [23]. Therefore, in this study, the soil particle volume fractal dimension D_v is chosen to evaluate the soil fractal features. The corresponding calculation model is expressed in Eq (1):

$$\lg\left[\frac{V(r < R_t)}{V_T}\right] = (3 - Dv)\lg\left(\frac{R_t}{R_{max}}\right)$$

(1)

Eq (1) is a double logarithmic curve function, in which V_T is the total volume of all soil particles, $V(r < R_t)$ indicates the total volume of the soil particles with sizes less than the radius R_t (mm), R_{max} is the radius of the maximal particle size, and D_v is the volume fractal dimension of the soil particle size distribution. In fact, the double logarithmic curves can be transformed into a linear function using $\lg\left[\frac{V(r<R_t)}{V_T}\right]$ as the y axis and $\lg\left(\frac{R_t}{R_{max}}\right)$ as the x axis. In this case, the slope (K) of the transferred linear function can be obtained by linearly

fitting to the equation $k = 3-D_v$, and the volume fractal dimension D_v can be calculated.

2.   Mass domain fractal dimension (D_{mi}) and volume domain fractal dimension (D_{vi}).

The domain fractal dimension was developed by Bittelli et al. (1999) [12], who categorized soil particle size distribution domains primarily into clay, silt, and sand domains and discovered that the relationship between the cumulative mass of soil particles and the soil particle size does not obey strict linear relationship. Instead, only certain soil particle size domains have an obviously linear relationship between the corresponding cumulative mass and soil particle size. As a result, the fractal dimension calculation model was introduced along with three corresponding mass domain fractal dimensions (D_{mi}), including D_{mclay}, D_{msilt}, and D_{msand} for clay, silt and sand domain, respectively. The model is expressed in Eq (2):

$$\begin{cases} \dfrac{M(r<R)}{M_T} = \left(\dfrac{R}{R_{L.upper}}\right)^{v} \\[2em] D_{mi} = 3-v \end{cases}$$

(2)

$$\begin{cases} \dfrac{V(r<R)}{V_T} = \left(\dfrac{R}{R_{L.upper}}\right)^{v} \\[2em] D_{vi} = 3-v \end{cases}$$

(3)

In Eq (2), $M (r<R)$ is the mass of the soil particles a radius less than R (mm) and M_T is the total mass of particles with a radius less than the upper size limit $(R_{L.upper})$ for the fractal behavior, which is determined by the measurement of specific soil samples. In fact, due to the existence of three particle size domains (clay, silt, and sand), there theoretically exists three upper sizes for soil particles of clay, silt, and sand. v is the constant exponent, and D_{mi} is the mass domain fractal dimension calculated individually for the clay, silt, and sand domains.

Based on Eq (2) and the theory of mass domain fractal dimension, Wang et al. (2007) [32] also reported a similar result to that from Bittelli et al. (1999) [12] with respect to the description of the PSD. Wang et al. (2007) [32] measured and analyzed the relationship between the cumulative volume of soil particles and the PSD of soil samples collected in two typical loess hilly-gullied watersheds located in Ansai county, Shaanxi province on the Loess

Plateau, China. However, unlike the mass domain fractal dimension method of describing the soil PSD utilized by Bittelli et al. (1999) [12], Wang et al. (2007) [32] mainly utilized the volume domain fractal dimension (D_{vi}) and Eq (3) instead of Eq (2) to describe the specific characteristics of the soil particle volume distribution.

In Eq (3), $V(r<R)$ is the volume of soil particles with a radius less than R (mm). V_T is the total volume of particles with a radius less than $R_{L.upper}$, and D_{vi} is the volume domain fractal dimension calculated for the clay, silt, and sand domains determined by the measurement of specific soil samples using *Malvern Mastersize 2000* laser diffusion and expressed as D_{vclay}, D_{vsilt}, and D_{vsand}, respectively. In this paper, the three volume domain fractal dimensions are simplified as D_{clay}, D_{silt} and D_{sand}.

3.   Volume domain fractal dimension based on USDA (Dvi(U)).

According to the United States Department of Agriculture (USDA) classification of soil particle size, soil particles are mainly divided into 7 partitions based on soil size. Based on the particle sizes, the clay domain is defined by particle sizes less than 0.002 mm, the silt domain is defined by particle sizes of 0.002~0.05 mm, and the sand domain particle size ranges from 0.05 mm to 2 mm. Therefore, the volume domain fractal dimension ($D_{vi}(U)$) is the representative of $D_{clay}(U)$, $D_{silt}(U)$, and $D_{sand}(U)$, which means that the upper size limit for fractal behavior ($R_{L.upper}$) described by Eq (2) is 0.002 mm, 0.05 mm, and 2 mm, respectively. Using the calculation of $D_{silt}(U)$ as an example, the calculation model can be further written as Eq (4):

$$D_{silt}(U) = 3 - \frac{\log(V_{silt} + V_{clay}) - \log(V_{clay})}{\log 50 - \log 2}$$

(4)

In Eq (4), the V_{silt} and V_{clay} are the volume fraction of silt and clay with particles sizes from 0.002 mm to 0.05 mm and less than 0.002 mm, respectively. The values 50 and 2 are the upper size limit for fractal behavior $R_{L.upper}$ of silt and clay, respectively, in the units of μm. Log is the natural logarithm.

To evaluate the differences between the two types of volume domain fractal dimension, as well as to assess the correlation and variability between the D_{vi} (D_{clay}, D_{silt}, and D_{sand}) $D_{vi}(U)$ ($D_{clay}(U)$, $D_{silt}(U)$, and $D_{sand}(U)$), two statistical metrics, the correlation coefficient (R) and root mean square error (RMSE), are chosen. The calculations of R and RMSE are as shown in Eqs (5) and (6), respectively. The calculations of D_{vi} (D_{clay}, D_{silt}, and D_{sand}) are based on the $D_{vi}(U)$ ($D_{clay}(U)$, $D_{silt}(U)$, and $D_{sand}(U)$) obtained from Eq (4).

$$R = \frac{cov(D_{vi}, D_{vi}(U))}{\sqrt{var(D_{vi})var(D_{vi}(U))}}$$

(5)

$$RMSE = \sqrt{\frac{\sum (D_{vi} - D_{vi}(U))^2}{N}}$$

(6)

In Eq (5), $cov(D_{vi,}D_{vi}(U))$ is the covariance of D_{vi} and $D_{vi}(U)$ and var(D_{vi}) and var($D_{vi}(U)$) are the variance of D_{vi} and $D_{vi}(U)$, respectively. R is the correlation coefficient (dimensionless). In Eq (6), N is the number of soil samples.

Statistical and other Analyses

Linear regression is used to fit the volume domain fractal dimension of the soil particle distribution and the soil properties in terms of soil organic matter and soil texture. The DUNCAN significant difference analysis is also carried out to compare the six main land-use types in terms of soil properties and fractal dimensions. All statistical analyses are conducted using SPSS17.0 software.

RESULTS AND DISCUSSION

Characteristics of Soil Particle Size Distribution by Land Use

According to the analysis of the relationship between soil particle size and cumulative volume percentage distribution in six main land-use types, including *Pinus massoniana Lamb, Robinia pseudoacacia L, Citrus reticulata Blanco, Zea mays L, Ipomoea batatas*) and*Setaria viridis*, the cumulative volume percentage of soil particles and the soil particle distribution over the entire particle distribution range do not exhibit a strict linear relationship for the purple soil areas. Instead, the linear relationships between the cumulative volume percentage of soil particles and the soil particle distribution are mainly grouped into volume domains, including clay, silt, and sand domains, which is similar to the results for America, Switzerland (Bittelli et al. (1999) [12]), and the Loess Plateau of China (Wang et al. (2007) [32]). The linear relationship between soil particle size and cumulative volume percentage is shown in Fig 2. (Fig 2 shows that the linear relationship between soil particle size and cumulative volume percentage of the six land-use types, and the upper size boundaries in the measured clay domain and silt domain of purple soil were approximately 5.74μm and 286.85μm.)

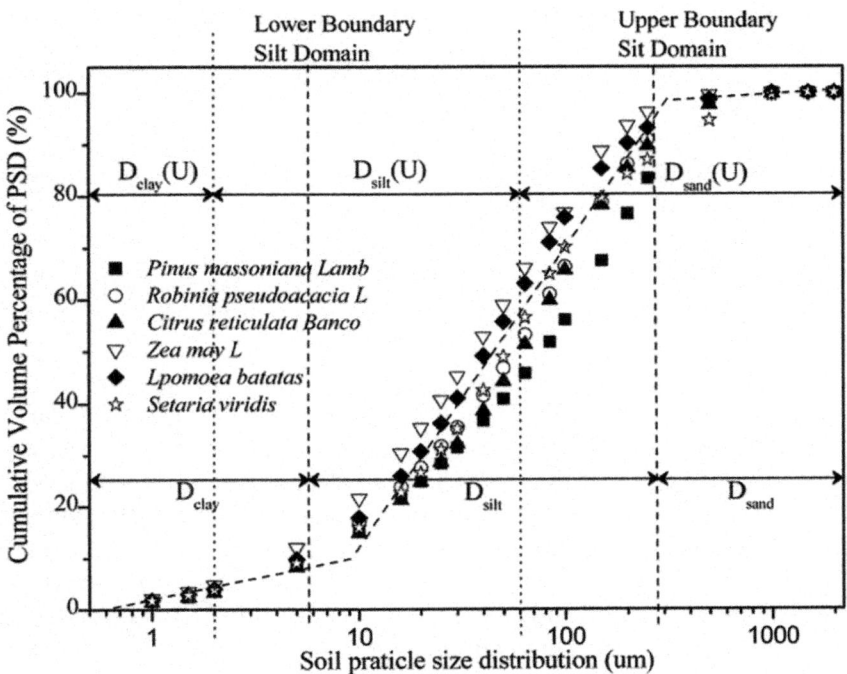

Figure 2. The characteristics of the soil size distribution by land-use type.

However, these values are not totally in agreement with the upper size limit for fractal behavior determined by the USDA classification standard, in which the upper size limits for clay, silt, and sand domains are 0.002 mm, 0.05 mm, and 2 mm, respectively. The volume domain of the upper size limit for the fractal behavior of purple soil in typical watersheds measured by laser diffusion is also shown in Fig 2.

Moreover, the measured volume domain of the upper size boundary of purple soil is also different from the corresponding value of the for loess soil obtained by laser diffusion by Wang et al. (2007) [32]. This is mostly due to the differences of the formation and development of the solid structure of loess soil and the differences in the parent materials.

In addition, in order to obtain the best fit of the relationship between the cumulative volume percentage and the purple soil particle size in measured volume domains (D_{clay}, D_{silt} and D_{sand}) (i.e., an R^2 value as calculated by Eq (5) closer to 1), the linear regression analysis method is utilized. The response characteristics of soil particle size on the cumulative volume percentages are also optimized. The fitting statistics and optimization results are shown in Table 2.

Table 2. Soil volume fractal dimensions and domain boundaries by land-use type.

Land-use types	D_v		Clay domain			Silt domain			Sand domain			Median diameter		Silt domain boundary μm			
														Upper boundary		Lower	
			D_{clay}			D_{silt}			D_{sand}			d_{50} μm				boundary	
	mean	R²	mean	R²	CV (%)	mean	R²	CV (%)	mean	R²	CV (%)	mean	CV (%)	mean	CV (%)	mean	CV (%)
Pinus massoniana Lamb	2.478a	0.95	1.910c	0.98	0.572	2.487ab	0.98	3.722	2.960ab	0.91	0.941	77.44a	35.331	286.966a	0.003	5.749a	0.003
Robinia pseudoacacia L	2.493a	0.92	1.921c	0.99	0.461	2.485ab	0.99	2.625	2.985a	0.90	0.460	58.06a	27.213	286.960a	0.001	5.743a	0.002
Citrus reticulata Blanco	2.470a	0.93	1.937bc	0.97	2.560	2.429b	0.96	4.857	2.950bc	0.92	1.354	61.91ab	41.095	286.958a	0.001	5.749a	0.003
Zea mays L	2.533a	0.90	1.933bc	0.98	0.348	2.531a	0.98	3.193	2.982ab	0.89	0.322	39.89b	39.142	286.852b	0.008	5.748a	0.017
Ipomoea batatas	2.501a	0.91	1.960ab	0.96	0.292	2.483ab	0.97	2.796	2.975ab	0.90	0.278	42.34b	24.370	286.877b	0.088	5.748a	0.020
Setaria viridis	2.489a	0.92	1.969a	0.96	0.981	2.462ab	0.98	1.151	2.925c	0.92	1.580	60.08ab	39.382	286.960a	0.001	5.749a	0.002

Footnotes: CV, coefficient of variation.

doi:10.1371/journal.pone.0122842.t002

As shown in Fig 2, differs from the USDA classification standard, the upper size boundaries in the measured clay and silt domain of purple soil for the six land-use types of the two typical small watersheds are approximately 5.74μm and 286.85μm, respectively. Moreover, Table 2 shows that, among the different land-use types, the volume domain upper size of clay ranges from 5.743μm to 5.749μm and the mean volume domain upper sizes for different land uses are not significantly different from each other (P<0.05). The upper size boundary in the measured silt domain of purple soil is 286.852~286.966 μm. Additionally, based on the determination of the volume fractal dimension calculated by Eq (1), the volume fractal dimension (D_v) and volume domain fractal dimension (D_{vi}) are calculated and the results are shown in Table 2. The results show that among the six various land-use types, the mean of D_v fluctuates between 2.470 and 2.533 without a significant difference (P<0.05) as it is compared with D_{vi}.

More specific information is also provided in Table 2. First, the calculated mean D_{clay} for all land uses using the laser diffusion instrument ranges from 1.910 to 1.969. The corresponding determination coefficient reaches 0.97, which indicates that the fitting of the relationship between the cumulative volume percentage of soil particles and soil particle distribution is reasonably good. The average D_{clay} values for different land use types have the following relationships: *Setaria viridis>Ipomoea batatas>Citrus reticulata Blanco>Zea mays L>Robinia pseudoacacia L>Pinus massoniana Lamb*. Wang et al. (2005) and Konert et al. (1997) reported that as it was compared with the volume contents of clay and silt measured by the pipette method [33], the values obtained using laser diffusion (specifically, by the application of *Malvern Mastersize 2000* laser diffusion) were lower and higher, respectively. And the D_{clay} for loess soil was even negative (Wang et al. (2005) [23]). However, in this paper, the value of D_{clay} for different land uses in two typical purple soil from the small watersheds is higher than what Wang et al. (2005) and Konert et al. (1997) reported. The average value of D_{clay} in our experiment reaches nearly 1.94, which is much greater than the value for loess soil. The D_{clay} differences between purple and loess soil are most likely due to the variability of the parent material, which is an important soil-forming factor affecting the soil physical and chemical properties. In fact, due to the higher weathering rate of the parent material of purple soil, the proportion of fine particles in the soil PSD becomes very high. The soil-forming periodicity of purple soil from parent material weathering to the mature soil formation is shorter than other soil type. Therefore, the average value of D_{clay} in purple soil is positive and even higher than loess soil.

According to the analysis and calculation of D_{silt} and D_{sand} by land use, the D_{silt} and D_{sand} of*Zea mays L* and *Ipomoea batatas*, *Robinia pseudoacacia L* and *Pinus massoniana Lamb*are all higher than the values of D_{silt} and D_{sand} in *Setaria viridis* and *Citrus reticulate Blanco*land uses located in the two typical purple soil watersheds. All of these values are similar to the results of the volume fractal dimension analysis on soil PSD under different land uses in the Loess Plateau by Wang et al. (2007) [32]. The higher D_{silt} and D_{sand} in *Zea mays L* and*Ipomoea batatas* land use indicate that fine soil particles are more common in the silt domain and sand domain. The higher values of D_{silt} and D_{sand} in *Zea mays L* and *Ipomoea batatas* land are most likely due to the human activities, such as tillage, cultivation, and fertilization. These activities are able to crush coarse soil particles into fine particles while improve the nutrient utilization of vegetation planted in these land uses.

The higher values of D_{silt} and D_{sand} for the soil samples collected in the *Robinia pseudoacacia L* and *Pinus massoniana Lamb* land uses indicate a higher proportion of fine particles in the silt and sand domains in typical forestland. The deep-root characteristics, which also play a pivotal role in improving soil structure, have greatly influenced the soil porosity and the quantity of soil microorganism. In addition to the effect of deep-root systems on the soil solid properties, the soil particle and nutrients, especially on the surface layer of the soil profile, are affected by the litter layer covering the forest soil. Therefore, distinct from reasons for the *Zea mays L* and *Ipomoea batatas* land, the plant physiological characteristics are most likely the main causes of the higher values of D_{silt} and D_{sand} for the forestland (*Robinia pseudoacacia L* and *Pinus massoniana Lamb*).

Moreover, regarding the lower D_{silt} and D_{sand} values calculated from the soil sampled with the *Setaria viridis* and *Citrus reticulate Blanco* land uses, *Setaria viridis* has a poor capability to conserve the soil structure due to its shallow-root physiological characteristics. In addition, large amounts of *Setaria viridis* in both of the two selected typical purple soil small watersheds are distributed on steep hillslopes and poor nutrient areas, which are able to influence the formation and the accumulation of the fine soil particles. There covers a thin layer of *Citrus reticulate Blanco*, which is an important plant introduced in the Grain-for-Green Project for both ecological and economic benefits in many purple soil watershed areas in southwestern China. The *Citrus reticulate Blanco* is able to protect many fine soil particles from being transported away while leaving coarser particles by the exceeded the runoff caused by the intensified precipitation over the threshold.

According to a comparison of the volume domain fractal dimension in different land uses, the value of D_{vi} followes the sequence of $D_{clay}<D_{silt}<D_{sand}$. This relationship is similar to the results from Bittelli et al. (1999). We also find that the average value of D_v is close to the average value of D_{silt}, and the absolute value of the difference between D_v and D_{silt} is 0.008~0.041. The higher determination coefficient ($0.89{\leq}R^2{\leq}0.99$) shown in Table 2indicates that the volume domain fractal dimension model is able to efficiently describe the volume distribution of the soil particle size. The determination coefficient of D_{silt} is higher than the determination coefficient of D_{clay}, and the D_{sand} is similar to the results from Bittelli et al. (1999) The reason to explain this fact was also given by Bittelli et al. (1999), in which the authors noted that based on the measurement of 19 soil samples from American and Switzerland, the classification and data acquisition for the silt and sand domains were most likely affected by the limitation of the experiments in terms of testing the soil PSD using the*Malvern Mastersize 2000*. More specifically, the error in the sand domain classification was mainly due to the processes of soil particle sieving; however, the limitations and accuracy of the silt domain classification were strongly affected by the application of laser diffusion technology, all of which may cause the difference between D_{clay}, D_{silt}, and D_{sand}.

Comparison of the Responses of D_{vi} and $D_{vi}(U)$ by Land Use

Based on Eqs (3)–(6), the relationship between D_{vi} and $D_{vi}(U)$ is shown by different classification standards in Fig 3. According to the D_{vi} and $D_{vi}(U)$ in the clay domain (Fig 3), the D_{clay} values are larger than the corresponding $D_{clay}(U)$ values (except for the D_{clay} of*Setaria viridis*, which is smaller than the corresponding $D_{clay}(U)$). The RMSE shown in Fig 3(a) is 0.030. Similarity, in the silt domain, the D_{silt} values of all six land-use types are consistently higher than the corresponding $D_{silt}(U)$ with a regularity that was also found forD_{sand} and $D_{sand}(U)$ for all six land-use types. Generally, the higher value of D_{vi} with regard to$D_{vi}(U)$ are affected by both of the classifications types and the volume domains standards. Moreover, according to the results from Bittelli et al. (1999), in which the RMSE of D_{silt} and$D_{silt}(U)$ were 0.090 for soil from America and Switzerland. This value is larger than that of D_{silt} and $D_{silt}(U)$ for soil from two typical purple soil small watersheds (0.058) in our experiments. The difference in the RMSE of D_{silt} and $D_{silt}(U)$ can be explained with the following two reasons. On one hand, the calculation of D_{vi} and $D_{vi}(U)$ by Bittelli et al. (1999) is mainly based on Eq (2), which essentially represents the mass domain fractal dimension (D_{mi}), rather than the volume domain fractal dimension (D_{vi}). In our experiments, Eq (3) is used to describe the volume fractal behavior of purple soil. On the other hand, the soil samples collected by Bittelli et al. (1999) were mainly derived from 7 parent materials, with the heterogeneity of soil properties. However, the parent material for the purple soil develops from gray-brown purple sand mudstone of the Shaximiao Formation (J_2S). Therefore, due to the relative homogeneity of the purple soil parent materials, the RMSE of D_{silt} and $D_{silt}(U)$ is higher than what Bittelli et al. (1999) reported. (Fig 3 shows the correlation analysis for Dvi and Dvi(U) for different land uses in typical purple soil small watershed, revealing that the correlation between Dclay and Dclay(U) of all six land-use types was not significant.)

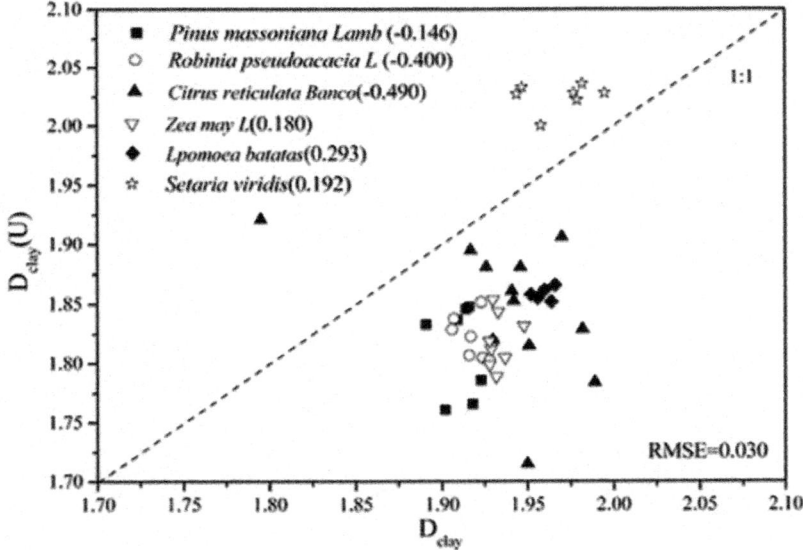

(A)

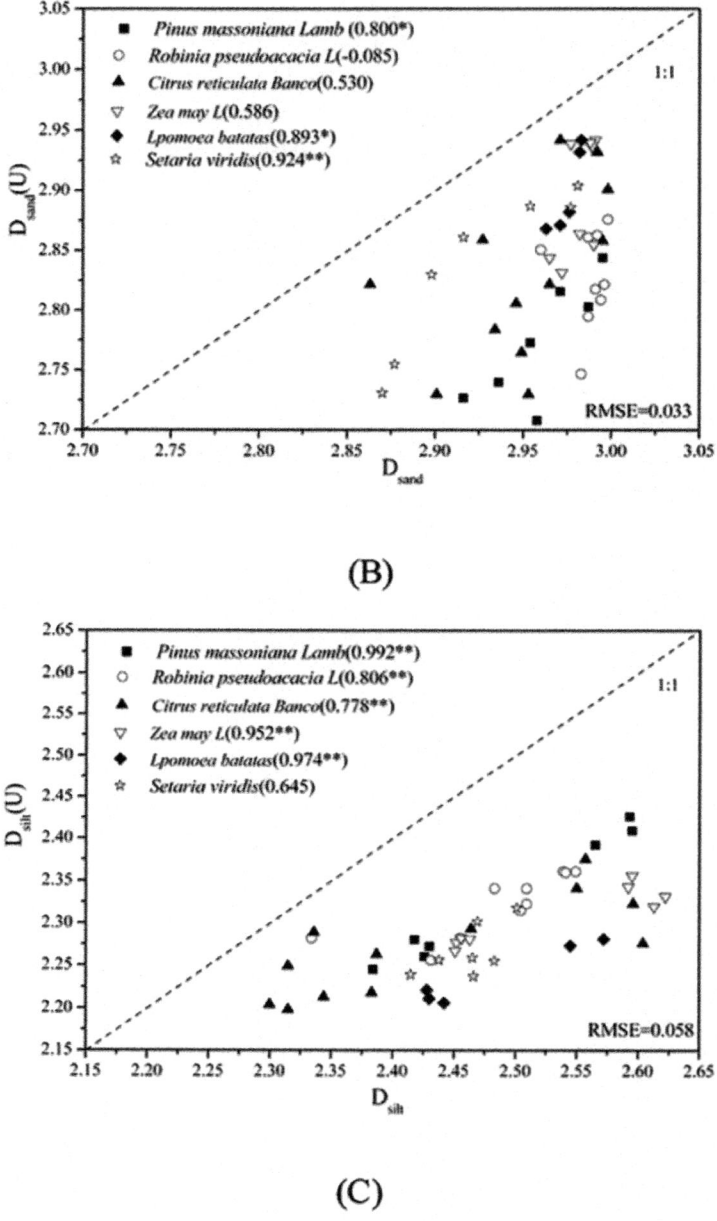

(B)

(C)

Figure 3. The comparison of the volume domain fractal dimension based on the measured D_{vi} and volume domain fractal dimension using USDA ($D_{vi}(U)$) land-use type classification. Footnote: the number in brackets represents the correlation coefficient; * indicates significant correlation ($P<0.05$), ** indicates significant correlation ($P<0.01$).

Fig 3 also shows the correlation analysis for D_{vi} and $D_{vi}(U)$ for different land uses in typical purple soil in the selected small watersheds. Generally, the correlations between D_{clay} and $D_{clay}(U)$ of all six land-use types are not significant. However, in the sand domain, the correlation between D_{sand} and $D_{sand}(U)$ for the *Pinus massoniana Lamb* (R = 0.800), *Ipomoea batatas* (R = 0.893), and *Setaria viridis* (R = 0.924) land uses are significant. In the silt domain, except for the correlation between D_{silt} and $D_{silt}(U)$ of *Setaria viridis* land use, the correlations between D_{silt} and $D_{silt}(U)$ of the other five land-use types are significant and positive. The correlation coefficient of *Pinus massoniana Lamb* land use is the highest (0.992) among all types, and the lowest correlation coefficient (0.778) appears in case of the *Citrus reticulate Blanco* land use.

Based on the significant correlation between D_{vi} and $D_{vi}(U)$ for different land uses in a typical purple soil in the selected small watershed, the reasons that the correlation between D_{silt} and $D_{silt}(U)$ is more significant than that between D_{sand} and $D_{sand}(U)$ and between D_{clay} and $D_{clay}(U)$ is listed below: Both of the ranges of D_{silt} and $D_{silt}(U)$ have wider soil PSD in the silt domain than the clay and sand domains. A wider soil PSD means that more soil particle size samples are included and contained in the wider distribution range, which is able to more efficiently and accurately represent all of the characteristics of the volume soil particle distribution relative to the clay and sand domains. Therefore, the correlation between D_{silt} and $D_{silt}(U)$ is more significant than that with clay and sand.

Response of Volume Domain Fractal Dimensions to the Soil Properties

The response of volume domain fractal dimensions to soil properties, i.e., soil texture and soil organic matter, is shown in Table 3 and Fig 4. The significance of the correlations between D_{silt}, $D_{silt}(U)$, D_{sand}, $D_{sand}(U)$, and D_{clay} and soil texture differs from each other, except for the insignificant correlation between $D_{clay}(U)$ and soil clay, silt, and sand contents. The correlations between the soil organic matter and the volume domain fractal dimensions are significant to different extents, which includes $D_{clay}(U)$, D_{silt}, $D_{silt}(U)$, and D_{sand} with the corresponding correlation coefficients of -0.420, 0.730, 0.621, and 0.606, respectively. (Fig 4, in the both typical purple soil small watersheds, except for the insignificant correlation between Dclay(U) and soil clay, silt, and sand contents, the significance of the correlations between Dsilt, Dsilt(U), Dsand, Dsand(U), and Dclay and soil texture differed. On the other hand, the correlativity of soil organic matter and volume domain fractal dimensions were significant to different extents.)

Table 3. The correlation analysis between Dvi (Dvi(U)) and soil texture and organic matter on a watershed scale.

Classification scale	Soil properties	D_v	Clay domains		Silt domains		Sand domains	
			D_{clay}	$D_{clay}(U)$	D_{silt}	$D_{silt}(U)$	D_{sand}	$D_{sand}(U)$
Watershed scale	Clay content	0.816**	—	—	0.913**	0.723**	0.604**	0.582**
	Silt content	0.735**	0.420**	—	0.829**	0.441**	0.606**	0.689**
	Sand content	-0.748**	-0.406**	—	-0.846**	-0.480**	-0.615**	-0.593**
	Organic matter	0.564**	—	-0.420**	0.730**	0.621**	0.606**	—

Footnotes:
**, significant correlation (P<0.01);
—, no significant correlate

doi:10.1371/journal.pone.0122842.t003

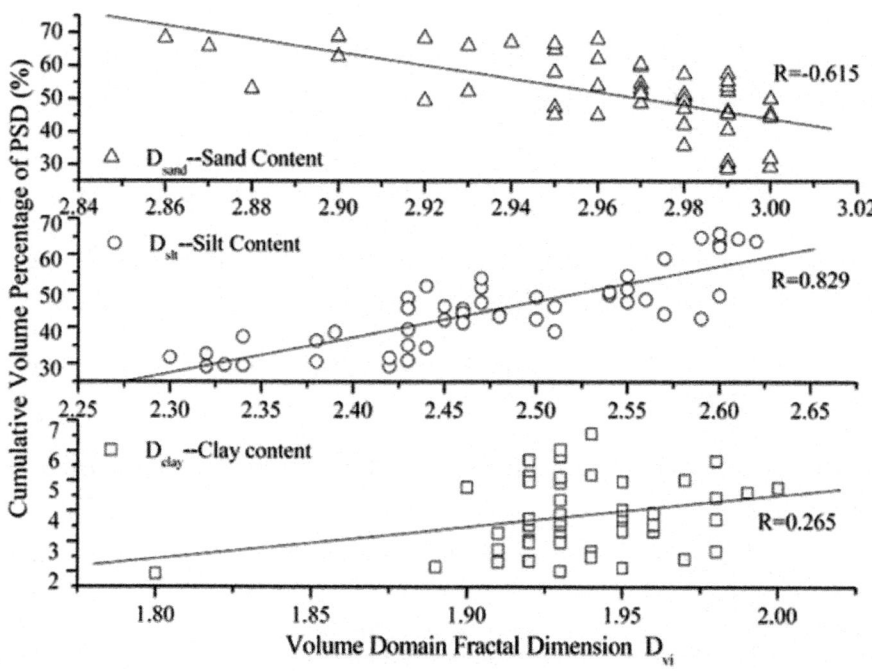

(A)

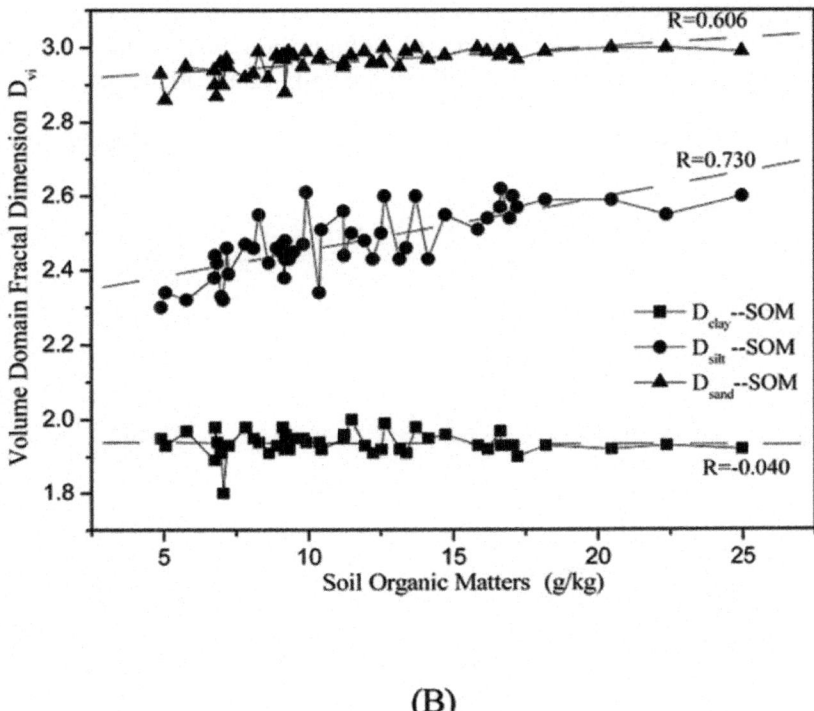

(B)

Figure 4. The correlation between volume domain fractal dimension based on measured D_{vi} and soil texture (A) and linear regression analysis for D_{vi} and soil organic matter (B) on a watershed scale.

The correlation between the soil properties in terms of soil texture as well as the organic matter and volume domain fractal dimension in terms of D_{vi} and $D_{vi}(U)$ is analyzed. The results are shown in Table 4. Specifically, the correlation analysis is conducted between the soil texture, including clay, silt and sand content for different land uses, and their corresponding volume domain fractal dimensions, including D_{clay}, D_{silt}, and D_{sand}, in the typical purple soil for the selected small watersheds. In addition, the correlation between the soil organic matter and volume (D_{vi} and $D_{vi}(U)$) is also analyzed (Table 4). Moreover, using the soil properties (soil texture and organic matter) for the *Citrus reticulate Blanco* land use as an example, these correlations between D_{vi} (D_{clay}, D_{silt} and D_{sand}) are also obtained. A linear regression analysis is carried out to fit the characteristics of volume domain fractal dimension and corresponding soil properties, such as soil texture and organic matter. The fitting results are shown in Fig 5. (Fig 5, analysis of the correlation between Dvi(Dclay, Dsilt and Dsand) and soil properties (soil texture and organic matter) in Citrus reticulate

Blanco land use, these correlations were also described using linear regression analysis to fit the characteristics of volume domain fractal dimension and corresponding soil properties (soil texture and organic matter).)

Table 4. The correlation analysis between D_{vi} and $D_{vi}(U)$ and soil texture and organic matter for six land-use types.

Land-use type	Soil properties	D_{vi}	Clay domains		Silt domains		Sand domains	
			D_{clay}	$D_{clay}(U)$	D_{silt}	$D_{silt}(U)$	D_{sand}	$D_{sand}(U)$
Pinus massoniana Lamb	Clay content	0.980**	—	-0.907**	0.989**	0.971**	0.874*	0.832*
	Silt content	0.953**	—	-0.805*	0.949**	0.922**	0.874*	—
	Sand content	-0.927**	—	0.927**	-0.956**	-0.914**	-0.912**	-0.861*
	Organic matter	0.935**	—	—	0.939**	0.897**	0.904**	0.778*
Robinia pseudoacacia L	Clay content	0.871**	—	-0.783**	0.873**	0.810**	—	—
	Silt content	0.895**	0.705*	-0.700*	0.921**	0.783**	0.720*	—
	Sand content	-0.901**	-0.693*	0.719*	-0.924**	0.795**	-0.712*	—
	Organic matter	0.862**	—	-0.715*	0.804**	0.761*	0.648*	—
Citrus reticulata Blanco	Clay content	0.910**	—	—	0.964**	0.863**	0.705*	0.614*
	Silt content	0.848**	—	—	0.953**	0.690*	0.753**	—
	Sand content	-0.860**	—	—	-0.961**	-0.712**	-0.754**	-0.581*
	Organic matter	0.702*	—	—	0.976**	0.645*	0.702*	—
Zea mays L	Clay content	0.986**	—	—	0.978**	0.899**	0.856**	—
	Silt content	0.972**	—	—	0.987**	0.949**	0.907**	—
	Sand content	-0.983**	—	—	-0.988**	-0.945**	-0.904**	—
	Organic matter	0.783*	—	—	0.781*	0.885**	0.766*	—
Ipomoea batatas	Clay content	—	—	—	—	—	—	—
	Silt content	—	—	0.906*	—	—	—	—
	Sand content	—	—	-0.894*	—	—	—	—
	Organic matter	—	—	—	—	—	—	—
Setaria viridis	Clay content	0.766*	—	—	0.928	—	0.810*	—
	Silt content	—	—	—	0.757*	—	0.826*	0.775*
	Sand content	—	—	—	-0.779*	—	-0.831*	-0.773*
	Organic matter	0.799*	—	—	0.890**	—	—	—

Footnotes:
*, significant correlation (P<0.05);
**, significant correlation (P<0.01);
—, no significant correlation.

doi:10.1371/journal.pone.0122842.t004

Table 4 shows that the D_{vi} and soil properties are significantly correlated, except for the cases of the *Ipomoea batatas* and *Setaria viridis* land uses. The correlations are especially significant for the *Robinia pseudoacacia L* and *Pinus massoniana Lamb* land uses. Moreover, Table 4 also shows that the correlation between D_{silt} and soil properties is significant for five of the land uses (*Pinus massoniana Lamb, Robinia pseudoacacia L, Citrus reticulata Blanco, Zea mays L, and Setaria viridis*), which is similar to the correlativity between D_{sand} and soil properties. However, the correlativity between D_{clay} and soil texture and organic matter is not significant for all land uses. Additionally, in the cases that the calculation of the $D_{vi}(U)$ is based on the USDA classification and standards. Table 4 also indicates that the correlation between $D_{clay}(U)$ and soil texture and organic matter is not significant, neither does the sand domain. However, in the silt domain, the correlations are significant for four land uses: *Pinus massoniana Lamb, Robinia pseudoacacia L, Citrus reticulata Blanco*, and *Zea mays L*. Table 4 also indicates that the correlation between sand content and all the volume domain fractal dimensions, including D_{vi} and $D_{vi}(U)$, is significant and negative correlated, which is similar to the results from Wu et al.(1993), Yang et al.(2008), and Wang et al(2005). [10,23–24].

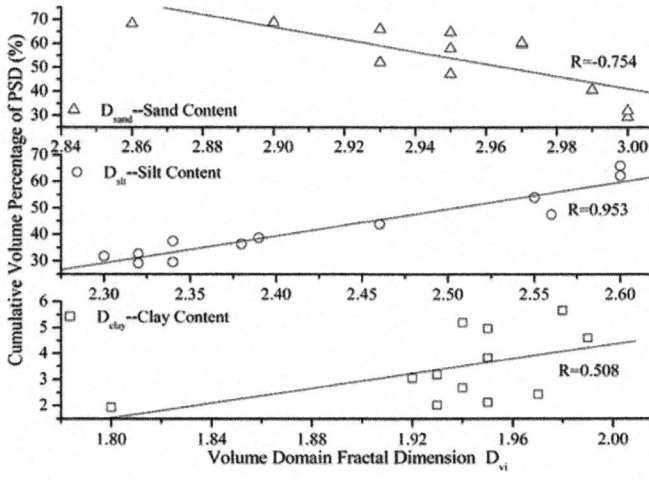

(A)

(B)

Figure 5. The correlation between volume domain fractal dimension based on the measured D_{vi} and soil texture (A) and linear regression analysis for D_{vi} and soil organic matter (B) for the *Citrus reticulata Blanco* land-use type.

Additionally, as it is shown in Table 3, the D_{silt}, $D_{silt}(U)$, D_{sand}, and $D_{sand}(U)$ have stronger correlativity with soil texture and show more obvious responses to the soil property, comparing with that in the clay domain (D_{clay}). The soil texture is ranked as the following:$D_{silt}>D_{silt}(U)>D_{sand}(U)>D_{sand}$. The responses of D_{clay} and D_{clay} to the soil organic matter in all the land-use types are weak due to the poor correlation. However, the D_{silt}, $D_{silt}(U)$, D_{sand}, and $D_{sand}(U)$ are positively correlated to varying extents of the soil organic matters. The intensity of the response of the volume domain fractal dimension to the soil organic matter has the following order: $D_{silt}>D_{silt}(U)>D_{sand}>D_{sand}(U)$.

In addition, regarding the response of volume domain fractal dimension to the soil properties in different land uses of the two typical purple soil small catchments, the six land uses are classified into two groups based on their functionalities: forestland, which is the representative of *Robinia pseudoacacia L*, *Pinus massoniana*, and *Citrus reticulata Blanco*, and farmland, which is the representative of *Zea mays L* and *Ipomoea batatas*. Based on classification of the land uses, Table 5 shows the response of the volume domain fractal dimension of forestland and farmland to the soil properties (soil texture as well as the organic matter) in terms of the correlation coefficients for different volume domain fractal dimensions and the soil texture and soil organic matter. Moreover, for the example of agricultural land, the relationship between D_{vi} and soil texture and organic matter are presented using a linear fitting analysis, as it is shown in Fig 6. (Fig 6, the relationship between Dvi and soil texture and organic matter were subjected to linear fitting analysis on the agricultural land.)

Table 5. The correlation analysis between D_{vi} and $D_{vi}(U)$ and soil texture and organic matter in the forestland and the agricultural land types.

Land-use type	Soil properties	D_v	Clay domains		Silt domains		Sand domains	
			D_{clay}	$D_{clay}(U)$	D_{silt}	$D_{silt}(U)$	D_{sand}	$D_{sand}(U)$
Forestland(P/R/C)	Clay content	0.924**	0.373*	-0.505**	0.921**	0.840**	0.674**	0.526**
	Silt content	0.829**	0.551**	-0.437**	0.844**	0.581**	0.681**	0.587**
	Sand content	-0.844**	-0.534**	0.461**	-0.864**	-0.625**	-0.695**	-0.604**
	Organic matter	0.714**	—	-0.498**	0.762**	0.708**	0.642**	—
Agricultural land(Z/L)	Clay content	0.891**	—	-0.638*	0.917**	0.828**	0.803**	—
	Silt content	0.887**	—	—	0.896**	0.703**	0.764**	—
	Sand content	-0.896**	—	—	-0.903**	-0.721**	-0.772**	—
	Organic matter	0.678*	—	—	0.687**	—	—	—

Footnotes:
**, significant correlation (P<0.01);
—, no significant correlation.
P, R, C, Z, and L represent *Pinus massoniana Lamb*, *Robinia pseudoacacia L*, *Citrus reticulata Blanco*, *Zea mays L*, and *Ipomoea batatas*, respectively.

doi:10.1371/journal.pone.0122842.t005

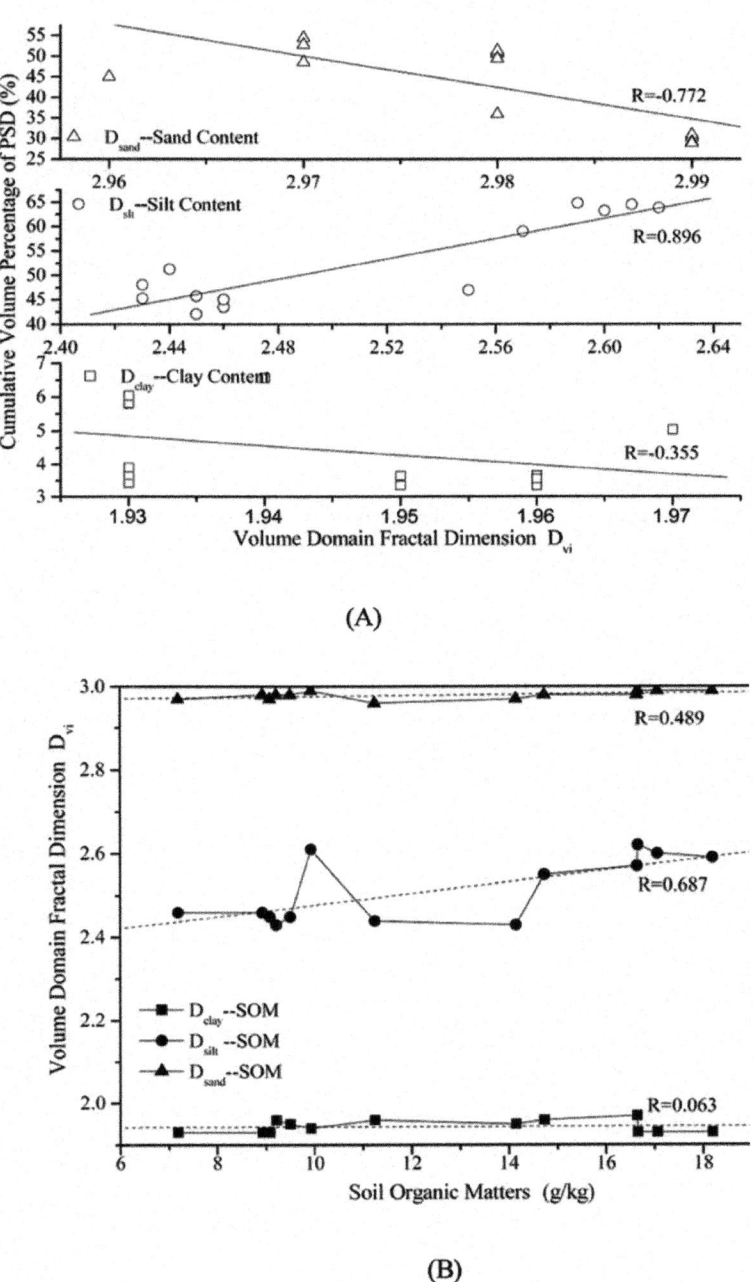

Figure 6. The correlation between volume domain fractal dimension based on the measured D_{vi} and soil texture (A) and linear regression analysis for D_{vi} and soil organic matter (B) for the agriculture land-use type.

From Table 4, the intensity of the response of the volume domain fractal dimension to the soil texture for the forestland has the following order:$D_{silt}>D_{silt}(U)>D_{sand}>D_{sand}(U)>D_{clay}>D_{clay}(U)$. The correlation coefficients between the volume domain fractal dimension and soil organic matter for $D_{silt}(U)$, D_{sand}, and $D_{clay}(U)$, are 0.708, 0.642, and -0.498, respectively. The largest correlation coefficient appears in the soil type of silt. In the agricultural land, including *Zea mays L* and *Ipomoea batatas*, only D_{silt}, $D_{silt}(U)$, and D_{sand} have significant correlations with the soil texture. However, only in forestland, D_{silt} has a significant correlation (R = 0.687) between volume domain fractal dimension and soil organic matter.

The influence of human activities, such as tillage, fertilization, and other soil management practices, on the response of the volume domain fractal dimensions to the soil properties is analyzed. To better illustrate the difference, we group the land-use types based on their intensity of exposure to human activities. *Pinus massoniana Lamb*, *Robinia pseudoacacia L* and *Setaria viridis* are categorized into the group I landuse, which is not affected by intensive human activities. The group II land use consists of *Zea mays L*, *Ipomoea batatas*, and *Citrus reticulate Blanco* land uses, all of which are intensively affected by human. The correlation test and a linear regression analysis are carried out. The results for both of the two groups are shown in Table 6 and Fig 7. (Fig 7, group I has a similar correlation between volume domain fractal dimensions, and the correlativity of Dsilt between soil organic matter was most significant in the corresponding groups.)

Table 6. The correlation analysis between D_{vi} and $D_{vi}(U)$ and soil texture and organic matter for group I and II land uses.

Land-use type	Soil properties	D_v	Clay domains		Silt domains		Sand domains	
			D_{clay}	$D_{clay}(U)$	D_{silt}	$D_{silt}(U)$	D_{sand}	$D_{sand}(U)$
Group I(P/R/S)	Clay content	0.908**	—	—	0.907**	0.777**	0.531**	0.519**
	Silt content	0.800**	0.413*	—	0.714**	0.468*	0.494*	0.683**
	Sand content	-0.814**	—	—	-0.807**	-0.542**	-0.561**	-0.698**
	Organic matter	0.708**	—	-0.617**	0.838**	0.806**	0.640*	—
Group II(C/Z/L)	Clay content	0.921**	—	-0.438*	0.943**	0.821**	0.690**	0.636**
	Silt content	0.889**	0.399*	-0.472*	0.944**	0.666**	0.757**	0.647**
	Sand content	-0.898**	—	0.472*	-0.949**	-0.686**	-0.755**	-0.650**
	Organic matter	0.725**	—	—	0.755**	0.505*	0.632**	0.573**

Footnotes:
**, significant correlation (P<0.01);
—, no significant correlation.
P, R, C, Z, L, and S represent *Pinus massoniana Lamb*, *Robinia pseudoacacia L*, *Citrus reticulata Blanco*, *Zea mays L*, *Ipomoea batatas*, and *Setaria viridis*, respectively.

doi:10.1371/journal.pone.0122842.t006

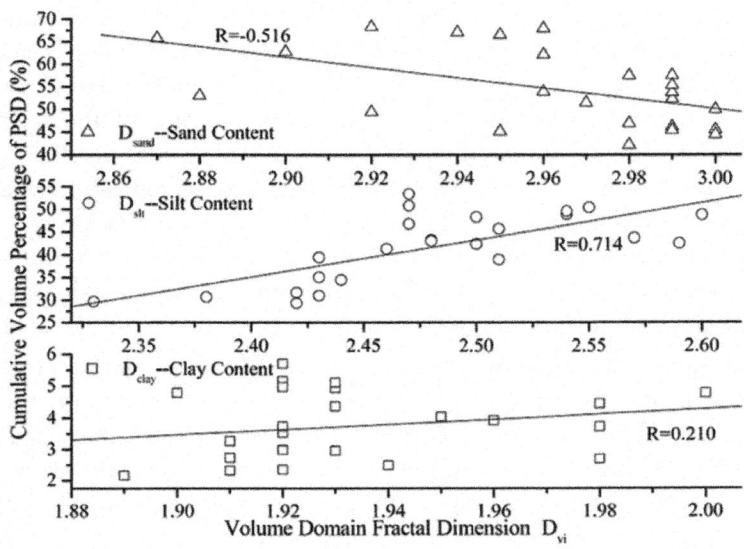

(A)

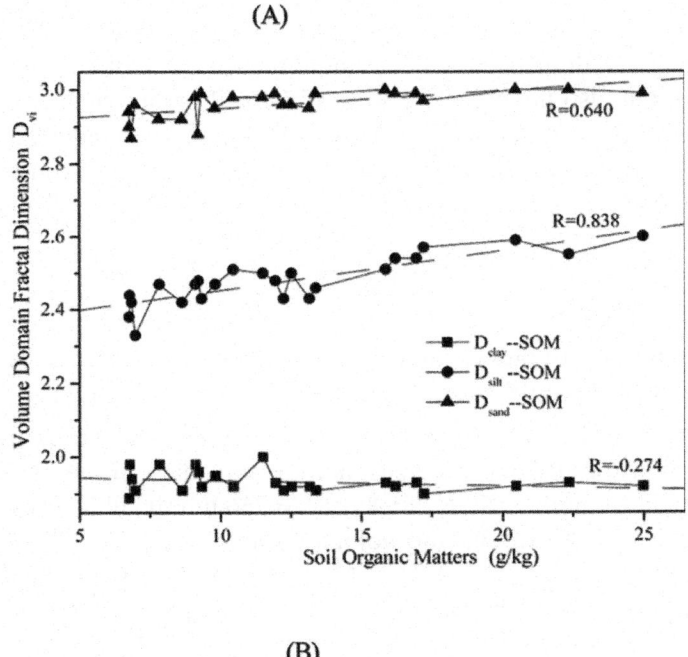

(B)

Figure 7. The correlation between the volume domain fractal dimension based on the measured D_{vi} and soil texture (A) and linear regression analysis for D_{vi} and soil organic matter (B) for group I land use.

As it is shown in Table 5, the correlation between D_{clay} and soil texture in group II is not significant. The other volume domain fractal dimensions, including D_{silt}, $D_{silt}(U)$, D_{sand}, $D_{sand}(U)$, and $D_{clay}(U)$, have different extents of significant correlations with soil texture. The intensity of the response of these volume domain fractal dimensions to the soil texture has the following order: $D_{silt} > D_{silt}(U) > D_{sand} > D_{sand}(U) > D_{clay}(U)$. In contrast, group I has a similar correlation between volume domain fractal dimensions, including D_{silt}, $D_{silt}(U)$, D_{sand}, and $D_{sand}(U)$, and soil texture. Moreover, in both group I and group II, the correlativity of D_{silt} between soil organic matter is the most significant. The correlation coefficients reach 0.838 and 0.755 for group I and II, respectively.

This method to group the land use types yields a correlation between D_{silt} and soil properties in terms of soil texture and organic matter, both of which are the key representatives of the indicators of soil quality [34]. This correlation is the most significant under the purple soil condition. It indicates that the volume domain fractal dimensions, especially D_{silt}, are able to be used as potential indicators of the soil texture and soil productivity. Moreover, given the fact that there is a significant positive correlativity between D_{silt} and the silt and clay content, as well as the fact that there exists a significant negative correlativity between D_{silt} and sand content, it is able to conclude that, the D_{silt}, as a volume domain fractal dimension, reflects not only the degree of fragmentation of soil particles but also the characteristics of the soil organic matter coupling with the soil particles in purple soil. Therefore, D_{silt} is more suitable than any other mass or volume domain fractal dimensions for describing and evaluating the characteristics of the relationship between soil texture, organic matter, and soil particles.

CONCLUSION

The conclusions of the response characteristics of soil fractal features by land use in a typical purple soil watershed are summarized below:

The volume domain upper size of the clay domain ranges from 5.743μm to 5.749μm for all land-use types, and the boundary of the upper size of the silt domain for purple soil is 286.852~286.966 μm. In all land-use types under the purple soil condition, the volume domain fractal dimensions have the following order: $D_{clay} < D_{silt} < D_{sand}$. Regarding the land uses, the values of D_{silt} and D_{sand} in *Citrus reticulate Blanco* and *Setaria viridis batatas* are smaller than the corresponding values in *Pinus massoniana Lamb*, *Robinia pseudoacacia L*, and *Ipomoea*. In addition, for all land-use types, all of the parameters in D_{vi} (D_{clay}, D_{silt}, and D_{sand}) are higher than the corresponding parameters in $D_{vi}(U)$ ($D_{clay}(U)$, $D_{silt}(U)$, and $D_{sand}(U)$). Moreover, regarding the response of the volume domain fractal dimension to the soil properties, the strengths of the correlation to the soil texture and soil organic matter have the following ranks: $D_{silt}>D_{silt}(U)>D_{sand}(U)>D_{sand}$ and $D_{silt}>D_{silt}(U)>D_{sand}>D_{sand}(U)$, respectively. Finally, due to the fact that D_{silt} has the most significant correlativity to the soil texture and organic matter among the various land uses of typical purple soil watersheds, it can be regarded as a potential indictor for evaluating the proportion of fine particles in PSD, as well as a key measurement for soil quality and productivity studies.

ACKNOWLEDGMENTS

We would like to thank Chongqing General Station of Soil and Water Conservation and Ecological Environment Monitoring of China for their help and support. The authors wish to thank Prof. Lei for his very careful English corrections and suggestions to the paper, which significantly improved the readability of the article.

AUTHOR CONTRIBUTIONS

Conceived and designed the experiments: XYC BLL LQD YHH JZ. Performed the experiments: BLL JZ LQD YHH. Analyzed the data: BLL JZ. Contributed reagents/materials/analysis tools: BLL JZ LQD YHH. Wrote the paper: BLL XYC. Improving the language with a professional terminology, compact sentence structure and exquisite words: TTY.

REFERENCES

1.   Mandelort BB. Fractals: form, chance and dimension. San Francisco: Freeman; 1977.

2.   Mandelort BB. The fractal geometry of nature. San Francisco: Freeman; 1982. pp. 45–256.

3.   Arya LM, Heitman JL, Thapa BB, Bowman DC. Predicting saturated hydraulic conductivity of golf course sands from particle size distribution. Soil Science Society of America Journal.2010; 74: 33–37. doi: 10.2136/sssaj2009.0022

4.   Falconer KJ. Fractal geometry: mathematical foundations and applications. Chichester: John wiley and sons; 1990.

5.   Rieu M, Sposito G. Fractal fragmentation, soil porosity and soil water properties: Application. Soil Science Society of America Journal. 1991; 55: 1231–1238. doi: 10.2136/sssaj1991.03615995005500050006x

6.   Miao CY, Ni JR, Borthwick AGL, Yang L. A preliminary estimate of human and natural contributions to the changes in water discharge and

sediment load in the Yellow River. Global and Planetary Change. 2011; 76(3–4): 196–205. doi: 10.1016/j.gloplacha.2011.01.008

7. Cheng XF, Shi XZ, Wang HJ. Fractal characteristics of particle of Arable Layers in Hilly Region of red soil. Scientia Geographica Sinica. 2003; 23(5): 617–621.

8. Scott WT, Stephen WW. Application of fractal mathematics to soil water retention estimation. Soil Science Society of America Journal.1989; 53: 987–996. doi: 10.2136/sssaj1989.03615995005300040001x

9. Scott WT, Stephen WW. Fractal scaling of soil particle size distributions: analysis and limitations. Soil Science Society of America Journal.1992; 56: 362–369. doi: 10.2136/sssaj1992.03615995005600020005x

10. Wu Q, Borkovec M, Sticher H. Study on particle-size distribution in soil. Soil Science Society of America Journal.1993; 57: 883–889. doi: 10.2136/sssaj1993.03615995005700040001x

11. Perfect E, Kay BD, Rasiah V. Multifractal method for soil aggregate fragmentation. Soil Science Society of America Journal.1993; 57: 896–900. doi: 10.2136/sssaj1993.03615995005700040003x

12. Bittelli M, Campbell GS, Flury M. Characterization of Particle-size distribution in soil with a fragmentation model. Soil Science Society of America Journal. 1999; 63: 782–788. doi: 10.2136/sssaj1999.634782x

13. Prosperinin N, Perugini D. Particle size distributions of some soil from the Umbria Region (Italy):Fractal analysis and numerical modeling. Geoderma.2008; 145: 185–195. doi: 10.1016/j.geoderma.2008.03.004

14. Posadas AND, Gimenenz D, Bittelli M, Vaz CMP, Flury M. Multifractal characteristics of soil particle size distribution. Soil Science Society of America Journal.2001; 65, 1361–1367. doi: 10.2136/sssaj2001.6551361x

15. Posadas AND, Gimenenz D, Quiroz R, Protz R. Multifractal characteristics of soil pore system. Soil Science Society of America Journal.2003; 67: 1361–1369. doi: 10.2136/sssaj2003.1361

16. Miao CY, Ni JR, Borthwick AGL. Recent changes of water discharge and sediment load in the Yellow River basin, China. Progress in Physical Geography.2010; 34(4):541–561. doi: 10.1177/0309133310369434

17. Caniego FJ, Espejo R, Martin MA, Jose FS. Multifractal scaling of soil spatial variability. Ecological Modeling.2005; 182: 291–303. doi: 10.1016/j.ecolmodel.2004.04.014

18. Segal E, Shouse PJ, Bradford SA, Skaggs TH, Corwin DL. Measuring particle size distribution using laser diffraction: implications for predicting soil hydraulic properties. Soil Science.2009; 174: 639–645. doi: 10.1097/ss.0b013e3181c2a928

19. Yang PL, Luo YP, Shi YC. The Description of the soil fractal characterization based on weight distribution of the particle size. Chinese Science Bulletin.1993; 38(20): 1896–1899.

20. Huang GH, Zhang WH. Fractal property of soil particle size distribution and its application. Acta Pedologica Sinica.2002; 4: 490–497. pmid:12557558

21. Huang GH, Zhang RD. Evaluation of soil water retention curve with the pore-soild fractal model. Geoderma.2005; 127: 52–61. doi: 10.1016/j. geoderma.2004.11.016

22. Martin MA, Montero E. Laser diffraction and multifractal analysis for the characterization of dry soil volume-size distributions. Soil and Tillage Research.2002; 64: 113–123. doi: 10.1016/s0167-1987(01)00249-5

23. Wang GL, Zhou SL, Zhao QG. Volume Fractal dimension of soil particles and its application to land use. Acta Pedologica Sinica.2005; 42(4): 546–550.

24. Yang JL, Li DC, Zhang GL. Comparison of mass and volume fractal dimensions of soil size distributions. Acta Pedologica Sinica. 2008; 45(3): 413–419.

25. Sun QH, Miao CY, Duan QY, Kong DX, Ye AZ, Di ZH, et al. Would the 'real' observed dataset stand up? A critical examination of eight observed gridded climate datasets for China. Environmental Research Letters.2014. doi: 10.1088/1748-9326/9/1/015001. pmid:25574186

26. Su YZ, Zhao HL, Zhang TH. Soil properties following cultivation and non-grazing of a semi-arid sandy grassland in northern China. Soil & Tillage Research.2004; 75: 27–36. doi: 10.1016/s0167-1987(03)00157-0

27. Zhao WZ, Liu ZM, Cheng GD. Fractal dimension of soil particle for sand desertification. Acta Pedologica Sinica.2002; 39(5): 877–881.

28. Wang D, Fu BJ, Zhao WW, Hu HF, Wang YF. Multifractal characteristics of soil particle size distribution under different land-use types on the Loess Plateau, China. Catena.2008; 72: 29–36. doi: 10.1016/j. catena.2007.03.019

29. Liu X, Zhang GC, Heathman GC, Wang YQ, Huang CH. Fractal features of soil particle size distribution as affected by plant communities in the forested region of Mountain Yimeng, China. Geoderma. 2009; 154: 123–130. doi: 10.1016/j.geoderma.2009.10.005

30. Zhang Z, Wei CF, Xie DT, Gao M, Zeng XB. Effect of land use patterns on soil aggregate stability in Sichuan Basin, China. Particuology.2008; 6: 157–166. doi: 10.1016/j.partic.2008.03.001

31. Zhang JC, Li HD, Lin J. Spatial variability of soil erodibility (K-factor) at a catchment scale in China. Acta Ecologica Sinica.2008; 28(5): 2199–2206.

32. Wang D, Fu BJ, Chen LD, Zhao WW, Wang YF. Fractal analysis on soil particle size distribution under different land-use types: a case study in the loess hilly areas of the Loess Plateau, China. Acta Ecologica Sinica.2007; 27(7): 3081–3089.

33. Dane JH, Topp GC. Soil Science Society of American Book Series NO.5 Part 4 Physical Method. SSSA Inc. Madison WI. 2002.

34. Dumanski J, Pieri C. Land quality indicators: research plan agriculture ecosystems and environment. Agriculture, Ecosystems & Environment.2000; 81: 93–102. doi: 10.1016/s0167-8809(00)00183-3

CITATION

CHAPTER 1

Wisley Moreira Farias, Geraldo Resende Boaventura, Éder de Souza Martins, Fabrício Bueno da Fonseca Cardoso, José Camapum de Carvalho and Edi Mendes Guimarães (2014). Chemical and Hydraulic Behavior of a Tropical Soil Compacted Submitted to the Flow of Gasoline Hydrocarbons, Environmental Risk Assessment of Soil Contamination, Dr. Maria C. Hernandez Soriano (Ed.), ISBN: 978-953-51-1235-8, InTech, DOI: 10.5772/57234.

CHAPTER 2

Rainer Schuhmann, Franz Königer, Katja Emmerich, Eduard Stefanescu and Markus Stacheder (2011). Determination of Hydraulic Conductivity Based on (Soil) - Moisture Content of Fine Grained Soils, Hydraulic Conductivity - Issues, Determination and Applications, Prof. Lakshmanan Elango (Ed.), ISBN: 978-953-307-288-3, InTech, DOI: 10.5772/20369.

CHAPTER 3

E. Logmo, G. Ngon, W. Samba, M. Mbog and J. Etame, "Geotechnical, Mineralogical and Chemical Characterization of the Missole II Clayey Materials of Douala Sub-Basin (Cameroon) for Construction Materials," *Open Journal of Civil Engineering*, Vol. 3 No. 2A, 2013, pp. 46-53. doi: 10.4236/ojce.2013.32A006.

CHAPTER 4

M. FALL, D. Sarr, M. Ba, E. Berbinau, J. Borel, M. Ndiaye and C. Kane, "Evolution of Lateritic Soils Geotechnical Parameters during a Multi-Cyclic OPM Compaction and Correlation with Road Traffic," *Geomaterials*, Vol. 1 No. 3, 2011, pp. 59-69. doi: 10.4236/gm.2011.13010.

CHAPTER 5

Santosh Kumar Sarkar, Paulo J.C. Favas, Dibyendu Rakshit and K.K. Satpathy (2014). Geochemical Speciation and Risk Assessment of Heavy Metals in Soils and Sediments, Environmental Risk Assessment of Soil Contamination, Dr. Maria C. Hernandez Soriano (Ed.), ISBN: 978-953-51-1235-8, InTech, DOI: 10.5772/57295.

CHAPTER 6

Melo, V., Uchôa, S., Senwo, Z. and Amorim, R. (2015) Phosphorus Adsorption of Some Brazilian Soils in Relations to Selected Soil Properties. *Open Journal of Soil Science*, **5**, 101-109. doi: 10.4236/ojss.2015.55010.

CHAPTER 7

Qissab, M. (2015) Flexural Behavior of Laterally Loaded Tapered Piles in Cohesive Soils. Open Journal of Civil Engineering, 5, 29-38. doi: 10.4236/ojce.2015.51004.

CHAPTER 8

Sall, O., Ba, M., Ndiaye, M., Sangare, D., Fall, M. and Thiam, A. (2015) Influence of Mechanical Properties of Concrete and Soil on Solicitations of Mat Foundation. *Open Journal of Civil Engineering*, 5, 249-260. doi:10.4236/ojce.2015.52025.

CHAPTER 9

Jian Li, Xiyong Wu, and Long Hou, "Physical, Mineralogical, and Micromorphological Properities of Expansive Soil Treated at Different Temperature," Journal of Nanomaterials, vol. 2014, Article ID 848740, 7 pages, 2014. doi:10.1155/2014/848740

CHAPTER 10

Ruiz S, Or D, Schymanski SJ (2015) Soil Penetration by Earthworms and Plant Roots—Mechanical Energetics of Bioturbation of Compacted Soils. PLoS ONE 10(6): e0128914. doi:10.1371/journal.pone.012891

CHAPTER 11

Luo B-l, Chen X-y, Ding L-q, Huang Y-h, Zhou J, Yang T-t (2015) Response Characteristics of Soil Fractal Features to Different Land Uses in Typical Purple Soil Watershed. PLoS ONE 10(4): e0122842. doi:10.1371/journal. pone.0122842

INDEX

A

Ampelisca abdita 112

B

bending moments 159, 175
biogeochemical weathering attributes
    130
biopolymers 194, 207
bioturbation 193, 205, 208, 215, 217,
    219, 220, 221
Brazilian soils 129, 130

C

cation exchange capacity (CEC) 3, 30
cationic exchange capacity (CEC) 15
CBR (Californian Bearing ration) 72
certified reference materials (CRM) 99
Citrus reticulate Blanco 233, 241, 245,
    247, 252, 254
concrete 159, 160, 162, 163, 174, 175

D

Derived equations 155
Differential scanning calorimetric (DSC)
    59
differential thermal (DT) 179, 183

dithionite-citrate bicarbonate (DCB) 131

E

earthworm biopore networks 193
earthworm hydroskeleton 193
earthworm hydrostatic skeleton 197
energy dispersive spectra (EDS) 58
European Community Bureau of Refer-
    ence (ECBR) 95
European Community (EC) 93

F

Filonenko-Borodich model 163
finite elements model (FEM) 208
Flexural behavior 145, 146
fluid flow 21

G

Gasoline 1, 15, 16, 20, 259
geotechnical engineering 71
global positioning system (GPS) 97
Gravel lateritic soils 71

H

Heavy metal pollution 87
Hierarchical cluster analysis (HCA) 103
hydraulic conductivity 3, 5, 6, 7, 10, 11,